From shark attacks to nuclear meltdow
demics, David Burrett Reid vividly sho
aren't always what we fear most. *Run*
reveals how our gut feelings about danger can mislead us, and
why understanding risk is essential not just for survival – but for
building a smarter, safer world.

Jim Clifton, Chairman, Gallup

I found this book lucid and well-rooted in history providing context for today's serious risks. It helped focus my attention on our
primary challenges.

Vint Cerf, internet pioneer

I have been studying risk for more than sixty years and have seen a
lot of books on the subject. *Running The Risk* is one of the best,
and I highly recommend it to anyone who wants a comprehensive,
readable, useful and up-to-date dive into this important topic.

Professor Paul Slovic, founder and
president of Decision Research

A genuinely fabulous book. An essential biography of risk covering
history, philosophy, psychology, religion and much more. My bet
is that this book will be talked about for years to come.

Professor Richard Clegg, chairman of the
Institute for the Public Understanding of Risk,
National University of Singapore

A must-read for anyone who wants to know about the risks we run in this highly complex world we're living in. This book provides wisdom and an optimistic vision for the future that is much needed in these challenging times.

Vincent Doumeizel, author of *The Seaweed Revolution* and senior adviser to the United Nations Global Compact

Entertaining and illuminating, this is odds-on to recalibrate your reasoning about every threat and opportunity. The only risk would be not to read it.

Quentin Cooper, science journalist and host of BBC Radio 4's *Material World* (2000–13)

With risk aversion all the rage in policy circles, and the antidote too often a glib lionizing of risk-taking's virtues, how refreshing to have a well-rounded book that draws on history in such a thought-provoking way. *Running the Risk* is packed with literary, philosophical, artistic, filmic, psychological, spiritual and mythical references. And the brilliant deep dive into the etymology of the word 'resilience' is just one of a myriad of reasons you should read this excellent book.

Claire Fox, Baroness Fox of Buckley, panellist on BBC Radio 4's *The Moral Maze* and Director of the Academy of Ideas

RUNNING THE RISK

RUNNING THE RISK

FROM SHARK ATTACKS TO NUCLEAR DISASTER – UNDERSTANDING LIFE'S BIGGEST RISKS AND HOW WE BUILD A SAFER FUTURE

David Burrett Reid

HERO, AN IMPRINT OF LEGEND TIMES GROUP LTD
51 Gower Street
London WC1E 6HJ
United Kingdom
www.hero-press.com

First published by Hero in 2025

© David Burrett Reid, 2025

The right of the author to be identified as the author and translator of
this work has been asserted in accordance with the Copyright, Designs
and Patents Act 1988. British Library Cataloguing in Publication Data
available.

Printed by Akcent Media, 5 The Quay, St Ives, Cambs, PE27 5AR

ISBN: 9781917163941

All the images in this volume are reprinted with permission or presumed
to be in the public domain. Every effort has been made to ascertain and
acknowledge their copyright status, but any error or oversight will be
rectified in subsequent printings.

All rights reserved. No part of this publication may be reproduced, stored
in or introduced into a retrieval system, or transmitted, in any form or
by any means (electronic, mechanical, photocopying, recording or otherwise), without the prior written permission of the publisher. This book is
sold subject to the condition that it shall not be resold, lent, hired out or
otherwise circulated without the express prior consent of the publisher.

Contents

It is better by noble boldness to run the risk
of being subject to half the evils we anticipate
than to remain in cowardly listlessness for fear
of what might happen.

– Herodotus, *The Histories*

RUNNING THE RISK

Chapter 1

A World of Risk

Did you risk your life today? The answer is almost certainly yes: we risk our lives every single day without even realising it. Getting in your car and going to work is one of the riskiest things you will ever do, although it might feel perfectly safe. When you ask people what they believe to be the greatest risk in their daily life, 'travelling by road' is consistently the top answer, and they're right. Road-related accidents are the leading cause of accidental death worldwide, with around 3,700 fatalities every day. Only a few natural causes of death, like heart disease or cancer, take more lives in a day. The workplace isn't much safer. Among the world's three billion workers, one in five reports they have been seriously injured at work at some point in their careers. The International Labour Organization estimates that nearly three million people die from occupational accidents and work-related diseases every year. Some of our most entrenched habits, like crossing the road or going to work, are so familiar they feel safe while they are, in fact, the most dangerous.

We are surrounded by hidden risks, but we tend to ignore them because we don't understand them or realize how big a threat they pose. Millions of people fear flying, campaign groups organize protests against nuclear power stations, and we often worry about the food we eat. The likelihood of being harmed by these risks is incredibly low, yet we believe they could be imminent. Commonplace risks like road traffic accidents rarely make it into national newspapers, but if one person is killed by a shark off the coast of Florida it creates headlines all around the world. On

3

average, only four people are killed by a shark every year. You're more likely to risk your life going to work than being eaten by a shark, yet shark attacks get much more attention and generate far more fear.

In the summer of 1975, the world was experiencing an extraordinary heatwave with record temperatures when the blockbuster film *Jaws* was released and caused widespread panic. People refused to go to the beach or go swimming; some even went as far as to hire boats and emulate the film's protagonists by hunting down sharks. But fear of sharks (or galeophobia) is not irrational: predatory fish are scary. Great whites, like the killer in *Jaws*, have mouths lined with up to 300 dagger-like teeth and a bite force of more than 4000 p.s.i. Many people who are afraid of sharks cite *Jaws* as the reason, having never actually seen one in the wild. The film tapped into powerful human emotions: fear of the unknown, the belief that the shark is a diabolical schemer, and the sense of dread invoked by the possibility of such a horrific death. Fifty years later, fear of sharks is deeply embedded in the public consciousness.

It isn't just films that influence our perception of risk; our beliefs are also shaped by the people around us, our culture or history, and what we read in the media or see on TV. Every coastal community has legends about people being attacked by sharks or other sea monsters, and they often appear as mortal enemies in novels or folk stories. One of the earliest records of a shark attack is from 1749, when Brook Watson, a fourteen-year-old cabin boy on the ship *Royal Consort*, dived into Havana harbour for a swim. As he floated, a shark sank its teeth into his leg, pulling him beneath the waves in a vicious, sustained attack that severed his right foot. A group of sailors managed to pull him free and he survived. Thirty years later, the incident was depicted by the American portrait painter John Singleton Copley in a monumental oil painting called *Watson and the Shark*, which now hangs in the National Gallery of Art in Washington, DC. Despite his injury, Watson went on to have a successful career as a merchant, becoming chairman of Lloyd's of London, a Member of Parliament, and the Lord Mayor of London.

John Singleton Copley, 1778, *Watson and the Shark*
(National Gallery of Art, Washington, DC)

Journalists say that news is anything that sells newspapers and, although they are relatively rare, shark attacks are big news. Being eaten by a shark is almost unheard of, but every few years, there is a gruesome example that generates sensational media headlines. In 2019, a British tourist was eaten by a shark while swimming off the coast of the Indian Ocean island of Réunion. He was missing for some time before a tiger shark was seen swimming near tourist beaches and was killed. A post-mortem examination found that its stomach contained a severed hand bearing the wedding ring of the missing man. The prominence of shark attacks in the media makes people believe they are much more frequent than they really are. On average, there are around seventy unprovoked shark attacks globally per year, and of these only four are fatal. The likelihood of being killed by a shark is roughly one in 3.7 million; you are much more likely to be hit by lightning, the odds of which are 300,000 to one. We shouldn't be afraid of sharks, but most people aren't aware of the probabilities involved. Even if they were, such abstract numbers are almost impossible to understand. It is well known that the brain is terrible at big numbers, especially when strong emotions are involved.

Being eaten by a shark is so terrible it causes panic, but we have got used to 1.3 million deaths on the road every year. Sharks cause much more worry but pose far less risk in our daily lives than travelling by road or going to work. Why do we accept some of the biggest risks around us yet worry about risks that are statistically highly improbable, like a plane crash or radiation leak from a nuclear power station? Humans aren't rational; they are ruled by emotions, biases, and personal experience. How we think and feel about risk has a huge influence on our behaviour and the actions we take to keep ourselves safe. Knowledge is the key. A better understanding of the risks around us could help us prioritize what we worry about, know what to do to protect our families or local community, and save millions of lives. So, what are the biggest risks we face every day, and what do we need to know about them?

Ninety Seconds to Midnight

In March 2023, against the backdrop of the Ukraine conflict, the Russian Federation announced its intention to 'share' nuclear weapons with Belarus. The first such agreement since the UN Treaty on the Non-Proliferation of Nuclear Weapons came into force in 1970, this move prompted Izumi Nakamitsu, the UN nuclear disarmament chief, to warn the Security Council that 'the risk such weapons will be used is higher today than at any time since the end of the Cold War'. She stressed that the war in Ukraine represented 'the most acute example of that risk'. As countries around the world returned to some kind of normal following years coping with the Covid-19 pandemic, suddenly, the war in Ukraine put fears of nuclear conflict at the forefront of public consciousness.

Anxiety that Putin might use nuclear weapons grew fast. Not long after the Russian invasion in February 2022, a Reuters/Ipsos poll showed that 58 per cent of Americans believed they were heading towards a nuclear war with Russia, and experts warned that the level of angst over nuclear conflict was greater than at any time since the Cuban missile crisis.[1] The Doomsday Clock, created by the Bulletin of Atomic Scientists in 1947 and a universally recognized indicator of the world's vulnerability, was reduced to ninety seconds to midnight as the war in Ukraine entered its second year – the closest to global catastrophe it had ever been. While the 2023 Bulletin also cited climate change, the energy crisis and disruptive technologies as factors in their decision, the threat of nuclear conflict was the foremost consideration.

Nuclear war has consistently been one of the most prominent risks faced by the planet as a whole throughout the second half of the twentieth century. At the height of the Cold War during the 1980s, the public's biggest fear was overwhelmingly that of a potential nuclear Armageddon, and there was a widespread belief that it might be only a button push away. In 1981, just a year after it was announced that Britain would allow the US to deploy cruise and Pershing missiles on UK soil, the British government published a short pamphlet called *Domestic Nuclear Shelters*, advising the public on how to build their own fallout shelter, where they could live for up to two weeks following a nuclear attack. Public information campaigns like this, together with protests by the Campaign for Nuclear Disarmament and the febrile international political environment, pushed nuclear to the top of people's concerns about their safety throughout the 1980s.

The closest the world has ever come to nuclear annihilation was during the Cuban missile crisis in 1962. Although short-lived, the crisis ushered in a new era of nuclear anxiety and had a profound impact on the national psyche. On the evening of 22nd October 1962,

John F. Kennedy informed the American public of the existence of missile sites on the island of Cuba, just ninety miles from mainland Florida. In his televised address, he said: 'We will not prematurely or unnecessarily risk the costs of worldwide nuclear war in which even the fruits of victory would be ashes in our mouth – but neither will we shrink from that risk at any time it must be faced.' Shock quickly turned to panic as the situation worsened, many believing that a devastating nuclear attack was imminent. There were runs on grocery stores as the public began to stockpile essential supplies such as food, water, toilet paper, flashlights and batteries. According to polls, thoughts about 'atomic bomb fallout' rose from 27 per cent in the spring of 1962 to 65 per cent during the crisis.[2]

Despite Nakamitsu's warning to the UN and the rising tensions created by the Ukraine war, 2023 wasn't like the 1960s or 1980s, when the nuclear threat was all-pervasive. While the threat might be one of the most severe faced by the world, it is too abstract and too remote to occupy many people's minds. Although we all face the same global risks, there is often a big difference between the issues that national governments and UN agencies worry about and the things the public is most concerned about. Like nuclear war in the 1960s and 1980s, the Covid-19 pandemic was an issue that dominated everyone's thoughts in all countries throughout the years from 2020 to 2022 – it was both a very personal and immediate risk to individuals and families as well as a crisis for governments, businesses, and all types of institution worldwide. In the aftermath of the Covid-19 pandemic, more immediate concerns over rising gas prices and the cost of groceries at the supermarket now dominate the public discourse in most countries. According to the *What Worries the World* survey, which tracks public opinion on the most important social and political issues across twenty-nine countries, inflation and the economy have been the top global concerns in 2022 and 2023.[3] Inflation is closely followed by crime and violence, poverty and social inequality, unemployment and financial or political corruption to make up the top five global worries.

Today's Crisis and Tomorrow's Catastrophes

What are the biggest risks we need to deal with worldwide right now and in the near future? Following almost twenty years of relative stability, the first years of the 2020s have heralded a particularly disruptive period in human history. As economies worldwide began to recover from the Covid-19 pandemic, the war in Ukraine has triggered crises in food and energy that will reverberate for years to come. We are now in the age of the 'polycrisis'. War, pandemics, climate change, inflation. Leaders around the world are facing multiple crises at the same time, with the effects compounding each other and our future trajectory looking more and more volatile.

Speaking to the broadcaster CBS in September 2022, President Joe Biden declared the coronavirus pandemic 'over' and made it clear that tackling inflation and controlling rising prices was now his number one priority. One crisis gave way to another, and the US government's risk register had changed. The World Economic Forum's *Global Risk Report*, which identifies the most severe risks the world is facing right now and up to a decade in the future, opens its 2023 report: 'As 2023 begins, the world is facing a set of risks that feel both wholly new and eerily familiar. We have seen a return of 'older' risks – inflation, cost-of-living crises, trade wars, capital outflows from emerging markets, widespread social unrest, geopolitical confrontation and the spectre of nuclear warfare – which few of this generation's business leaders and public policy-makers have experienced.'[4] According to the WEF report, the most significant risk facing society now and for the next few years is the cost-of-living crisis – driven by the global economic situation and exacerbated by the war in Ukraine and related energy crisis. Like Covid-19, it is a challenge faced equally by individuals, communities, governments and world leaders everywhere.

The cost-of-living crisis is a complex risk that has far-reaching consequences. Soaring food and fuel costs mean many people must choose between paying their bills or feeding their families. It is the most pressing worry for a staggering 93 per cent of Europeans as families across the EU struggle to make ends meet.[5] While it affects all demographics, it is most deeply felt by the poorest and most vulnerable groups and is gradually pushing more people into poverty. In an EU-wide survey, an estimated 8 per cent of the EU population (roughly 35 million people) could not keep their homes adequately warm in 2020, increasing the risk of developing mould — one of the leading causes of respiratory disease and asthma. The Institute of Health Equity estimates that 10–15 per cent of new childhood asthma cases in Europe are attributable to damp and mould.[6] So, the cost-of-living crisis forces people to make difficult choices about what they can afford, which can seriously affect their family's health and well-being. The medical journal *The Lancet* called the cost-of-living crisis the 'next major public health concern for both adults and children'.[7]

The 100 million people in Europe who suffer from chronic kidney disease are among the worst affected by rising electricity prices, which make it hard for many to afford the cost of their home dialysis machines. To make matters worse, because of their condition, they also feel the cold more acutely and so are more dependent on heating. The charity Kidney Care UK has reported that 44 per cent (32 per cent of whom are on dialysis) have missed meals due to the cost-of-living crisis, an example of the hard choices faced by some of the most vulnerable people. The situation has caused an explosion in food insecurity, defined by the WHO as 'people being able to access sufficient, safe, and nutritious food that meets their dietary needs', with women and families with children bearing the brunt of the situation. In a 2022 survey of six countries, French anti-poverty NGO Secours Populaire found that nearly half (48 per cent) of parents had cut back on their own food to feed their children.[8]

So what about the future? The *Global Risk Report* is based on a survey of the perceptions of 1,200 experts from academia, business, government, the international community and civil society. In the top ten risks it highlights some are 'societal', such as cost of living, social cohesion and forced migration; many are environmental risks, such as natural disasters and extreme weather; and one is technological: the rapidly increasing threat of cybercrime. In the 2025 Report, the number one risk was considered to be state-based armed conflict and new risks, such as the dangers posed by Artificial Intelligence, misinformation and disinformation, have now overtaken concerns about the cost-of-living crisis. However, the long-term risks it predicts are different. Over a ten-year horizon, the biggest risks are dominated by our potential failure to adapt to the effects of climate change and biodiversity loss.

On Friday, 14th October 2022, two young women, Phoebe Plummer and Anna Holland, walked into room 43 of the National Gallery in London and threw tomato soup over one of the most famous paintings in the world, one of Vincent van Gogh's *Sunflowers*, which has an estimated value in excess of £80 million. They then removed their jackets to reveal Just Stop Oil T-shirts before glueing themselves to the wall beneath the artwork. 'What is worth more, art or life?' they said. 'The cost-of-living crisis is part of the cost of oil crisis, fuel is unaffordable to millions of cold, hungry families. They can't even afford to heat a tin of soup.' They were part of the Just Stop Oil campaign group, which released a press statement shortly afterwards claiming that their actions were 'in response to the government's inaction on both the cost-of-living crisis and the climate crisis'. A few years earlier, Paralympic medallist James Brown climbed on top of a British Airways plane at London City Airport and filmed himself clinging to the fuselage at the end of a week of climate protests across the UK by Extinction Rebellion that saw over one thousand people arrested. This incident was just a few weeks after millions took to the streets of the world's biggest cities for the largest climate protest in history.

Public concern over climate change had been rapidly growing over the past ten years but took on a new sense of urgency when schoolchildren around the world started to protest. After Greta Thunberg went on a solo strike at her school in Stockholm, the 'school strikes' movement quickly spread, prompting strikes and demonstrations by schoolchildren in over 150 countries. A study published in 2021, which surveyed 10,000 children and young people aged sixteen to twenty-five in ten countries, found that 59 per cent were very or extremely worried about climate change (and 84 per cent were at least moderately worried).[9] These increasingly extreme protests are being caused by a growing sense of frustration at the lack of progress by the international community in slowing down the rate of climate change despite thirty years of warnings. According to the Intergovernmental Panel on Climate Change (IPCC), the chance of breaching the 1.5-degree target by 2030 stands at 50 per cent, while current commitments made by the G7 private sector suggest we will see an increase of 2.7 degrees by mid-century.

The failure to address climate change first entered the top rankings of the *Global Risks Report* in 2011, over a decade ago. Like the threat of nuclear war in the 1960s and 1980s, it has become the most pervasive and enduring risk in the public consciousness in recent years, with 'climate anxiety' replacing 'nuclear anxiety' among young people. Looking towards 2033, the report's experts clearly state that environmental risks dominate the long-term outlook. It puts failure of climate change mitigation and failure of climate change adaptation as the two most severe risks over the next decade, the first time that mitigation and adaptation have been included separately. It highlights the need to reduce greenhouse gases in the atmosphere, either by reducing emissions (from fossil fuels used for electricity, heat or transport) or by increasing the 'sinks' that remove these gases from the atmosphere and store them (such as seaweed, the oceans, forests), and to protect people from the adverse effects of life in a changing climate, for example reducing risks from sea-level rise, extreme weather or food insecurity. In the top ten risks by 2033, climate change is followed by other environmental issues, including natural disasters, extreme weather, biodiversity loss and the potential collapse of ecosystems worldwide. Cybercrime and the increasing threat from AI and other emerging technologies are also mentioned.

These risks are very real and already have a devastating effect worldwide. On 25th August 2022, Pakistan declared a state of emergency.

Following a severe heatwave, the country experienced heavier-than-usual monsoon rains that caused some of the deadliest flooding the world has ever seen. Thirty-three million people were left without shelter, many displaced and without food. Senator Sherry Rehman, Pakistan's minister for climate change, announced that 'literally one-third of Pakistan is under water. It's all one big ocean, there is no dry land to pump the water out... it's a crisis of unimaginable proportions.' It lasted from June until October, causing 1,739 deaths – 647 of which were children – and $14.9 billion in damage. Rehman called it 'a climate-induced humanitarian disaster'. A gradually warming world is causing an increase in the number and severity of heatwaves, extreme rainfall and storms or hurricanes.

The Bills of Mortality

> The fear of death is the most unjustified of all fears,
> for there's no risk of accident for someone who's dead.
>
> – Albert Einstein

Captain John Graunt, described by John Aubrey in his *Brief Lives* as 'very ingeniouse' and 'a pleasant facetious companion',[10] is hailed as being the forefather of both epidemiology and demographics for his work trying to understand what caused people to die in the seventeenth century. London was a vast city in 1660, with a population of around 350,000, making it one of the largest in Europe. But it was very crowded, living conditions were poor, and water from the river was dirty, so disease spread rapidly. Deaths across London were reported to the parish clerk every week and published on a Thursday as a kind of early warning system for plague epidemics, even though it was still a few years until the Great Plague would wipe out nearly 15 per cent of the population of London (over 65,000 people) in 1665.

Extracting all the useful data from the *Bills of Mortality*, Graunt published tables of statistics showing the frequency of deaths due to different causes, made estimates about the size of the population, calculated birth and death rates and gave insights into the spread of certain diseases. Among the causes of death included in his 'Table of Casualties' were commonplace conditions such as gout and consumption, strange maladies like 'King's Evil', 'frighted' and 'itch', and accidents such as drowning, which are still an issue today.

He pioneered public health statistics, and the *Natural and Political Observations Made upon the Bills of Mortality*, published in 1662, was the first published work of epidemiology. It is filled with firsts: he was the first to report that more boys than girls are born, offered the first reasoned estimate of the population of London and invented the 'life table'. Now a widespread tool used by actuaries, life tables show the probability that someone of a particular age will die before their next birthday. Global mortality statistics are helpful in understanding the most life-threatening risks. Graunt was also an early risk communicator. He wanted to allay unwarranted anxiety about certain risks that were feared far out of proportion to their likelihood of occurring. He produced the first hard evidence of how few deaths actually occurred from causes that terrified people in seventeenth-century London.

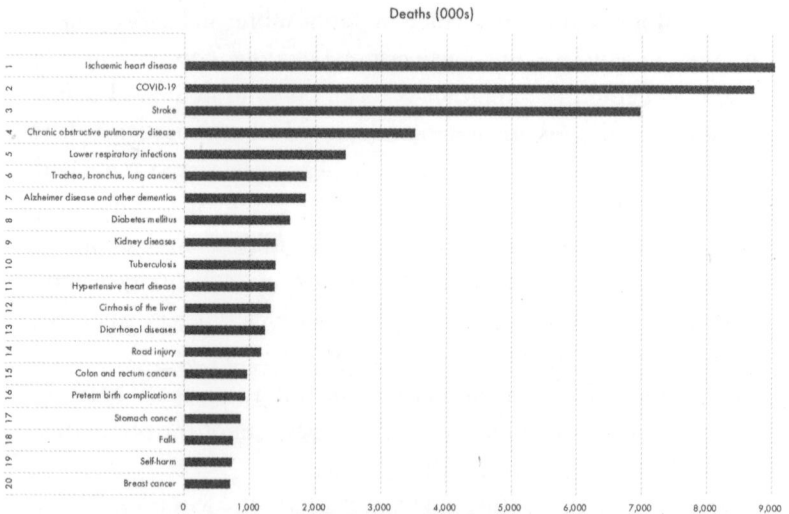

Estimated global deaths by cause, global health estimates 2021
(WHO, published 2024)

Graunt knew how much dread some of these strange diseases sweeping across London invoked – probably the biggest risks people worry about are those that could be fatal, either for them or for their loved ones. He believed the best way to alleviate that concern was to better understand the different causes and articulate them numerically, with solid evidence. Three hundred and sixty years later, the major causes of death haven't changed much. Mainly affecting older people, heart disease and cancer are still the leading causes of death globally, and of the fifty-six million people who die each year, half are aged seventy or older. These non-communicable diseases are the most common cause among high-income countries, while infectious diseases, malnutrition, HIV and malaria are common causes in low- and middle-income countries. While there isn't much we can do about ageing, understanding the underlying risk factors can be powerful information that *should* change our behaviour and help us avoid dying before our time. According to the Institute of Health Metrics and Evaluation Global Burden of Disease Study in 2019, the leading risk factors for premature death globally include high blood pressure, smoking, obesity, high blood sugar and air pollution – many of which we can control or at least avoid.

'Preventable injuries' are the fifth most common cause of death worldwide, with approximately 3.2 million deaths per year, and road traffic accidents are by far the biggest subcategory. Every year, around 1.2 million people have their lives cut short as a result of a motor vehicle crash, and between 20 and 50 million more are injured, many incurring permanent disabilities. Ninety-three per cent of these happen in low- and middle-income countries, and it is the leading cause of death for children and young adults aged 5–29 years old. This has led the United Nations General Assembly to set the ambitious target of halving the number of deaths and injuries from road traffic crashes by 2030.

The risk of having an accident is something we experience every day, and it can almost always be prevented if we take the right actions, either as individuals or as a community. The most common preventable injuries are falls, drowning, fire, poisoning and mechanical or work-related injuries. Every hour of every day, forty people die from drowning. In 2014, the World Health Organization (WHO) produced a report highlighting it as a hugely neglected public health issue. For every country in the world, drowning is in the top ten killers for children, while in some countries such as Bangladesh, it is the number one cause of death in children under the age of 15.[11] Yet it could be easily prevented by installing barriers, providing safe places for young children away from water, teaching children basic swimming, training rescuers, strengthening public awareness and building better resilience to flooding. As the WHO points out, the daily death toll from drowning continues its 'quiet rise' because the risk is not well known and fully appreciated by local communities and national governments.

While many commonplace accidents are overlooked, we are terrified of others. On 15th January 2009, Christopher Butler was sitting at his desk on the 33rd floor of the Nickelodeon offices in Midtown, New York, when he heard something strange and a co-worker pointed out a plane.

'We turned around. And sure enough, there was a plane coming in slow and level, and it touched down in the Hudson River... Obviously something was seriously wrong but it wasn't evident from the manner in which it was flying. It was just touching down. It seemed oddly uneventful.'

Butler was describing the emergency landing of US Airways Flight 1549, later called the Miracle on the Hudson, in which all 155 passengers were safely rescued. A regular scheduled domestic flight bound for Charlotte, North Carolina, it took off from New York's La Guardia airport normally, but early in its climb, it hit a massive flock of Canada geese so dense it blocked the pilot's windscreen. Almost immediately, both engines shut down, and Captain Chelsey Sullenberger – affectionately known as Sully – took control of the aircraft. Quickly realising they couldn't make it back to La Guardia, the only option was a water landing on the Hudson and evacuating the passengers by boat. Despite water landings being rare and dangerous, there was remarkably little damage to the plane and few serious injuries. You can even visit the aircraft, which has been preserved and is on display in the Carolinas Aviation Museum (now named the Sullenberger Aviation Museum).

So, how safe is flying? Accidents like bird strikes don't happen very often, and aircraft manufacturers have since made modifications to limit the impact of similar events. Over one-third of all people experience some form of anxiety when it comes to flying, and one in six people have a significant fear that prevents them from going on a plane. Some are surprised at how irrational this fear is when all the data shows that it is by far the safest way to travel – you are more likely to be killed by food poisoning or die from falling off a ladder than you are from flying. In 2022, there were only six fatal air accidents, claiming 174 lives, making it one of the safest years for air travel in the past ten years. On average, plane incidents cause death only once every 20 million flights, and, even when a crash does occur, 98.7 per cent do not result in a fatality. But such fears are easy to understand given that aeroplane disasters are massive global news stories, and the physics of flying are not as easy to understand (or are less visible) than the engineering involved in driving a car. And yet a fear of flying might actually kill you. In the wake of the 9/11 terrorist attacks, vast numbers of Americans switched from flying to driving, a much more dangerous way to travel long distances, resulting in an extra 1,595 people dying in road traffic accidents in the US in the year immediately after 9/11.[12] Fear and dread can make us do things that counter-intuitively put us at even greater risk.

What Is Risk?

Risk is usually thought of as the possibility of something bad happening or that something bad *might* happen. Whether through numerical probability or just by listening to our gut instincts, it's a concept we use to help us understand the world around us and make decisions about our lives. It is a way of predicting what the future might have in store so we can be better prepared, keep ourselves safe, or even take advantage of potential opportunities.

It isn't all life or death. Risk relates to any kind of negative consequence and can be very personal – it is present in our love lives and relationships, where there is always the possibility of being embarrassed, hurt, or betrayed. When he was twenty-nine years old, Vincent van Gogh wrote to his brother Theo, debating the risks of marrying his model Christine given his state of penury: 'Fishermen know that the sea is dangerous and the storm terrible, but they have never found these dangers sufficient to keep them ashore. They leave that philosophy to those who like it. Let the storm rise, the night descend – which is worse, danger or the fear of danger? Personally, I prefer reality, the danger itself.'[13] Like his art, Van Gogh's approach to personal risk, whether in the intimate sphere or his financial affairs, was based on his emotional and psychological response to the world – he opted for bold actions, like the bold colours and expressive images in his work.

The International Organization for Standardization defines risk vaguely as 'the effect of uncertainty on objectives', but for most of us, risks are usually synonymous with threats or potential harm. They can be about any kind of bad experience – as a whole, risk can be seen as anything that has a negative impact on something we care about, whether that is life, property or the world around us. There are many different types of risk. Some affect us personally, such as physical harm or financial loss; others have an impact on our community, our country or the natural environment. Some we can control, such as the food we eat or the transport we choose, and others, like a dirty bomb attack or tornado, are 'acts of God'. Many professions have formalized risk by putting a number on it and attempting to calculate the magnitude of a particular risk by looking at the probability and the severity of the potential outcome. Financial advisers look at the relationship between risk (the possible downsides of an investment) versus return and consider the likelihood of something happening and the potential cost involved; environmental professionals look at hazards and exposure; while medics look at the effects of certain risky behaviours on your health. Risk and safety are two sides of the same coin. The only difference is that risk is about the future (the possibility of harm), while safety is about the here and now, the things we do to protect ourselves from those potential dangers.

How Safe Do You Feel?

The travel writer Norman Lewis, who inspired Graham Greene to visit Saigon and write *The Quiet American*, described Cambodians in his 1951 book *The Dragon Apparent* as a gracious and endearing people: '[they] were Buddhist, and therefore in essence gentle, tolerant, and addicted to pleasures and satisfactions of a discriminating kind'.[14] Cambodians, most of whom are Buddhist, believe in animism and a rich supernatural world. They have a strong belief in ghosts, leaving spirit boxes next to their front doors filled with food and drink for wandering spirits who are hungry and thirsty to appease them and hopefully encourage them to keep wandering and not enter the house. According to their faith, 'hungry ghosts' are the spirits of people who have died violent, untimely or unnatural deaths. Given Cambodia's recent gruesome history, they have good reason for believing they are surrounded by these unhappy phantoms.

Roland Joffé's 1984 film *The Killing Fields* brought the genocide committed by the brutal Khmer Rouge regime led by dictator Pol Pot from 1975 to 1979 to the attention of the world for the first time. The film's title was coined by one of its subjects, Dith Pran, a Cambodian photojournalist who was captured after covering the fall of the capital, Phnom Penh, to the communists and later escaped. The Killing Fields are sites across Cambodia where over one million people were killed and buried after the end of the five-year-long civil war. At least 20,000 are scattered across the countryside, holding anywhere from a few bodies to hundreds, serving as burial grounds for Khmer Rouge victims as well as execution sites. As recently as 2007, researchers were still discovering previously unknown Killing Field sites, including one at Sre Leav, where 200 graves have been dug up and bones scattered throughout the woods by villagers hunting for jewellery. Buddhist monks led local villagers in prayers over the graves to try to make amends to the disrupted spirits. Though the pain of the past may have faded, the bones and the ghosts of Khmer Rouge victims still terrify many rural people. In the Cambodian countryside, they say, 'Don't go out at night. Ghosts,' 'Always live with other people. Ghosts,' or, 'Did you hear the crying last night in the quiet? Ghosts.' Each year, Terry McCoy, a Peace Corps volunteer working in Kompong Thom province, asks high-school classes to rank their greatest fears: 'Despite nagging concerns for HIV, cancer and leeches, ghosts always creep to the top of the list. At first, I found this amusing. Now, I get it.' McCoy realized that phasmophobia (the fear of ghosts) is embedded deep in the national psyche and has a powerful influence over people's behaviours. How we think about the risks around us and what provokes our fears is rarely based on hard evidence but rather stems directly from our national history, beliefs and culture. When trying to understand the public perception of risk and how people respond, context is everything.

So far, we have looked at what leading global experts can tell us about the risks around us and what we can learn from statistics about some of the most deadly risks. But do we really need to be told? We live with risk every day of our lives – crossing the road, sitting next to someone on the subway, going out alone at night – whether consciously or subliminally, we are constantly assessing the risks around us and making choices about our behaviour. Most of these day-to-day decisions are determined by quick, intuitive, 'gut' feelings rather than by any deliberate risk assessment. Based on our experience and values, these risk feelings are a powerful force, a kind of early warning system that can act as a guide to keep us safe. However, as we saw with the drop in people taking flights in the US after the 9/11 attacks, misperceptions of risk can sometimes have dangerous consequences. So, how safe do people in different countries feel, and what do they worry about the most?

Until recently, no comprehensive worldwide data existed about the public perception of risk. First carried out in 2019, the Lloyd's Register Foundation *World Risk Poll*, conducted by the global polling agency Gallup, was the first significant study to give a complete picture of how the world's citizens see risk and safety. Gallup interviewed over 150,000 people in 142 different countries. Much of the data they collected was from people who had never been surveyed before or from countries where little or no official figures exist yet where reported risks are often high. It gives an insight into how people feel in almost every country in the world, including some of the most remote and challenging places, where danger, death and the threat of injury are an everyday part of life.

Carried out just before the Covid-19 pandemic, the 2019 poll asked people how safe they felt compared to five years ago. The vast majority (72 per cent) of people worldwide said they felt as safe as or even safer than they did five years ago. However, Lebanon, Hong Kong, Afghanistan and Venezuela stood out, with the overwhelming majority of their citizens feeling less safe, mainly due to political turmoil or armed conflict. In Lebanon, large anti-government protests had erupted in October 2019, with hundreds of thousands of Lebanese taking to the streets; in Hong Kong, anti-China riots were just beginning; and in Afghanistan, there had been a surge in Taliban attacks. Conversely, people in mainland China and Rwanda felt much safer than they had five years earlier, most probably linked to the buoyant economic climate in both countries. After the 1994 genocide, Rwanda experienced twenty years of relative stability and economic growth, and before Covid-19, China was still experiencing the tail-end of a massive economic boom. Conflict, violence and financial prosperity all have a big impact on how safe people feel.

The 2019 *World Risk Poll* also asked people to name the greatest sources of risk to their safety in their daily lives. Road-related risks came top, especially in high-income countries, closely followed by crime and violence, which was most marked in Latin American countries, the Caribbean and South Africa. Health-related risks were cited more in regions with ageing populations and weak healthcare systems. All the data in the poll was collected before Covid-19, giving a snapshot of a pre-pandemic world before so much changed. The next poll was carried out in 2021. This time, the number of people who felt 'less safe' had increased slightly by 4 per cent globally but by over 10 per cent in areas hardest hit by Covid, including Southeast Asia, which had the world's highest death toll from the delta variant, and North America, which experienced record numbers of Covid-related deaths in early 2021. Seven per cent of people worldwide named Covid-19 as the greatest source of risk to their safety, making it the fourth most commonly named threat overall, behind road-related accidents, crime and violence, and non-Covid health issues. Despite their fear of ghosts, Cambodians told the 2021 *World Risk Poll* that the greatest source of risk to their safety was from more mundane, earthly sources: financial problems, Covid-19 or other health concerns. They also said they worry most about the potential of road traffic accidents and severe weather to cause them harm.

Taking Risks

> To dare is to lose one's footing momentarily. To not dare is to lose oneself.
>
> — Søren Kierkegaard

For some people, risk is good: an opportunity, something to seek out or embrace, a satisfying challenge to overcome. Indeed, in many languages, risk-taking is synonymous with adventurousness, courage, audacity, daring and bravery. In business, risk is essential for innovation and growth. In science, some of the most significant discoveries, such as the invention of the MRI scanner or the discovery of DNA's double helix, were from research considered adventurous, high-risk or even slightly eccentric at the time. However, in dangerous sports and exploration, the risks are unusually high, and the reward is harder to understand.

Every year, around 500 people attempt to climb Mount Everest, the highest mountain on Earth, with a peak at 8,849 metres above sea level. The year 2023 was the deadliest on record, with seventeen deaths, primarily due to altitude sickness, exhaustion or inexperience. The dangers faced by climbers attempting the summit are vast, including the risk of avalanches, falling rocks and ice, hypothermia, severe fatigue and health problems associated with extremely low oxygen. The death rate is approximately one per cent, which is pretty good for the 'eight-thousanders'. The most dangerous mountain in the world is the 8,091-metre main peak of the Annapurna massif, where around one-third of the teams who attempt the summit are lost. Despite this, mountain climbing has been one of the fastest-growing leisure activities of the last century, with an estimated ten million people going mountaineering annually in America alone. Why are these frightening, beautiful, icy landscapes so attractive, and why are so many climbers willing to risk their lives to reach a summit? George Mallory, whose death on the shoulder of Everest in 1924 shocked Britain, famously said, 'Because it is there.' But he also offered more insight into the existentialist mindset:

If you cannot understand that there is something in man which responds to the challenge of this mountain and goes out to meet it, that the struggle is the struggle of life itself upward and forever upward, then you won't see why we go. What we get from this adventure is just sheer joy. And joy is, after all, the end of life.

Three centuries ago, risking your life to climb a mountain would have been unthinkable. Only heroes and madmen would attempt such a thing, like Hannibal's crossing of the Alps in 218 BC, to surprise the Romans during the Second Punic War. Everyday citizens avoided them at all costs, believing them to be the dwelling place of demons, dragons and bandits. In the second half of the 1700s, during the Age of Enlightenment, people began to travel to mountain areas and even try to climb them – a new appreciation of the splendour of mountain ranges was born. A growing passion for seeking out knowledge was the driving force behind some of the first expeditions: collecting rocks, recording new terrain and performing experiments at altitude. In 1786, the summit of Mont Blanc was reached, and two years later, James Hutton's *Theory of the Earth* was published, heralding the birth of geology as a science. The beauty of these landscapes soon exerted its influence on the imagination of poets and artists as well, who captured the wonder of this icy realm – what John Ruskin called the 'endless perspicuity of space; the unfatigued veracity of eternal light'. Some of the mountains described by these artists became famous, and the first to reach the summit became heroes: Charles Barrington's ascent of the Eiger, Sir Edward Wymper reaching the summit of the Matterhorn, and Edmund Hillary's conquest of Everest were all front-page news across the world. This romanticism echoed down through the years to become the 'Boy's Own' adventure stories of derring-do read by millions of school children, enticing successive generations to take up mountain climbing. Perhaps the risk is worth it when experiencing beauty or achieving glory is the reward.

In his brilliant book *Mountains of the Mind*, the writer and climber Robert Macfarlane compares the chamois hunters of the Alps, who risk their lives every day chasing their quarry over steep slopes and glaciers, with the modern mountain climber. Of course, risk wasn't optional for the hunter – it came with the job. But the climber seeks it out:

> This is the great shift which has taken place in the history of risk. Risk has always been taken, but for a long time it was taken with some ulterior purpose in mind: scientific advancement, personal glory, financial gain. About two-and-a-half centuries ago, however, fear started to become fashionable for its own sake. Risk, it was realised, brought its own reward: the sense of physical exhilaration and elation which we would now attribute to the effects of adrenaline. And so risk-taking – the deliberate inducement of fear – became desirable: became a commodity.[15]

The word 'risk' can mean different things to different people and in other languages. While most people worldwide say the word 'risk' suggests danger, 45 per cent of people in central Asia associate risk with opportunity, which might explain why Singapore is ranked as one of the top three places in the world to start a new business. The oil-rich countries of the Middle East are also very positive towards risk, with the majority (57 per cent) of people in the United Arab Emirates thinking of risk as opportunity. It is not surprising, therefore, that Dubai has 4,777 tech start-ups. Entrepreneurs know that, while it brings uncertainty, taking risks generates possibilities for unexpected growth, fuels innovation, and leads to financial success if your gamble pays off. Frederick W. Smith, the founder of the world's biggest transportation company, FedEx, is now one of the world's wealthiest businessmen, but the success of his company was entirely based on a huge risk he took early in its history. In 1973, the company started offering services to twenty-five cities, transporting small documents and packages on their fleet of fourteen small jets. A week's fuel cost $24,000, and on one occasion, with only $5,000 left in the bank, the company was denied a crucial business loan. Smith took everything they had to Las Vegas, gambled it all at the blackjack table, and won $27,000, enough to keep the company afloat for one more week.

Fred Smith's gamble is an extreme example of survival, but many businesses have prospered from a high-risk, high-reward strategy, especially in the world of technology and innovation. Risk is inherent in every entrepreneur's new venture. Mark Zuckerberg, the founder and CEO of Facebook (now Meta), says: 'The biggest risk is not taking any risk. In a world that's changing really quickly, the only strategy that is guaranteed to fail is not taking risks.' Innovation is essential to the growth of most businesses, young or old, whether inventing a ground-breaking new product or finding creative solutions to existing problems. It involves taking risks, sometimes big risks and companies that take risks are usually more agile and adaptable, while risk-averse businesses often stagnate and fail. Drew Huston, a twenty-four-year-old MIT student, invented the file-sharing and backup service Dropbox because he was tired of carrying around USB sticks and emailing himself documents he wanted to share across different computers. He realized this was a problem shared by millions of people and founded the start-up that later became Dropbox. Apple co-founder Steve Jobs personally let Huston know that he was coming to take over the Dropbox market with his iCloud service, but Huston refused to sell to Apple: a considerable risk, as Apple could push them out of the market and make them obsolete in just a few years. It paid off, and Dropbox is now worth over $9 billion, while Huston's net worth is estimated to be over $2 billion.

Risk Power

In a rapidly changing world, risk is not fixed but constantly evolving. As Jonathan Swift put it, 'There is nothing constant in this world but inconsistency'. Rising temperatures due to climate change have dramatically increased the threat of severe weather events, such as tropical storms, drought and catastrophic flooding, which cause food and water shortages among some of the poorest people on the planet. Increasing technological and social complexity introduces new dangers we haven't even thought of before, such as cybercrime, the weaponisation of artificial intelligence and the misuse of our personal data. Human population expansion is increasing our demands on natural resources and, with it, our impact on fragile ecosystems and biodiversity, which could cause habitat loss or even collapse. Our world is also more interconnected than ever before, meaning risks are more global. The Covid-19 pandemic demonstrated that virus outbreaks spread like wildfire across the globe. Russia's invasion of Ukraine in 2022 showed that supply chains for essentials such as food and fuel are increasingly complex and interdependent.

From nuclear weapons and road traffic to floods caused by climate change and deadly pandemics, we live our lives surrounded by hundreds of different risks. We have learned to accept risk and even to tolerate it. Risk can cause worry and anxiety, but being oblivious to it can be even more dangerous. But how much risk is too risky? And how safe is safe enough? The more we understand the risks that we live with, the better our decisions. It can help us decide what food to buy or avoid, whether to invest in flood-proofing for our house, what to do in an emergency, whether to take a medical scan or what life-saving skills to teach our children.

Understanding risk can be a powerful tool that reduces anxiety and helps us lead safer, healthier, happier lives. To do that, we need to know more about the risks around us, how people have dealt with them in the past, and what we can do about future events. Risk knowledge is power. This book examines how communities all around the world have conquered risk throughout history and what we can learn from this to safeguard our families, make our society safer, and build greater resilience to cope with seen or unforeseen events.

Chapter 2

A Gift from God

We live in an uncertain world, but ancient civilizations often lacked the protections we take for granted today: laws, government, health-care, emergency aid or even a police force. With an average life expectancy of between twenty-five and twenty-eight years in ancient Greece or Rome, life was fragile and dangerous. Before the consolations of organized religion or probability theory, how did ordinary people assess risk and cope with uncertainty in their everyday lives?

The history of Greece is filled with accounts of warfare and violence. At the time, brutal displays like gladiator fights or battles with fierce animals were considered entertainment. Fourth-century Greece was a time of major upheaval – civil war was rife, and society was in a state of constant change or crisis. In the fifty years that followed the Peloponnesian War, Sparta, Thebes and Athens all fought to win a dominant position in the Greek world while contending with the pressing threat from Persia to the east. Despite numerous battles, it always ended in a stalemate until the emergence of a new power, the Kingdom of Macedonia, led by Alexander the Great. Out of this maelstrom came a school of self-help philosophy that taught Greek citizens that, no matter how challenging our lives, we can still thrive, even when the world is unpredictable and troubled. Developed by the philosopher Zeno in Athens, stoicism was a predecessor of the 'Keep Calm and Carry On' advice posters created in 1939 at the outbreak of the Second World War and still popular today. The stoics developed a mental fortitude that equipped them with the resilience needed to survive the most challenging times and the uncertainty of everyday life in ancient Greece. The famous stoic philosopher Epictetus taught that 'it is not events that disturb people, it is their judgements concerning them'. He believed stoicism was a way of life, not just a theoretical philosophy. Its central tenet was that external events are beyond our control, so we should accept whatever happens calmly and dispassionately. There is no point fighting against circumstances we can't control or getting attached to an outcome we have no power over.

Stoicism is a pragmatic approach to life that still resonates today and can be detected in many religions, popular psychology, and the teachings of self-help gurus in the media. While it says we should disregard things outside our control, it also makes it clear that we can master our own reactions and behaviour. The stoics recommended meeting every challenge with justice, self-control and reason. The Roman Emperor Marcus Aurelius, who was a famous Stoic, wrote in his *Meditations*:

If you are doing your proper duty, let it not matter to you whether you are cold or warm, whether you are sleepy or well-slept, whether men speak badly or well of you, even whether you are on the point of death or doing something else: because even this, the act in which we die, is one of the acts of life, and so here too it suffices to 'make the best move you can'.

The citizens of the ancient world had to contend with plague, famine, and other natural disasters, and they had little actual knowledge of the causes of these calamities or any warning or protection. The future was a mystery to them, and misfortune could be sudden and swift. Thucydides gives one of the most famous accounts of an early pandemic in his history of the Peloponnesian War, which started in 430 BC. In it, he writes: 'a pestilence of such extent and mortality was nowhere remembered'. It killed nearly one-third of the Athenian population during the second year of the war. In the year 165, during the reign of the last of the 'Five Good Emperors', Marcus Aurelius, the Antonine Plague swept across the Roman

Empire. Most likely a variant of smallpox, it originated in China and spread quickly westward along the Silk Road and was carried to the heart of the empire by trading ships and soldiers heading for Rome. The pandemic ravaged the city of Rome, and the contemporary historian Dio Cassius estimated over 2,000 deaths per day occurred in Rome alone. All told, it has been suggested that somewhere between one-quarter and one-third of the entire population of the Roman Empire fell victim to the pandemic, with a total death toll in the region of 60–70 million. It had a devastating impact on the Roman military machine, weakening their ability to defend the boundaries of the empire, while the economy took a steep decline due to a shortage of workers, the associated fall in productivity, and its impact on tax revenues. In *The History of the Decline and Fall of the Roman Empire*, Edward Gibbon writes: 'Pestilence and famine contributed to fill up the measure of the calamities of Rome.' Many subsequent historians have argued that the Antonine Plague should be seen as the starting point for the decline of the Roman Empire.

The angel of death striking a door during the plague of Rome. Reproduction of a wood engraving by P. Noël after J. Delaunay (Wellcome Collection)

Italy is one of the most seismically active places on Earth, so the ancient Romans would also have been very familiar with earthquakes which have been continually reshaping the area for millennia. The Greek city state of Helike disappeared without a trace after an earthquake and subsequent tsunami destroyed it in 373 BC.[1] The city and a huge space beneath it sank into the Earth and was covered over by the sea. All the inhabitants perished and ten Spartan ships anchored in the harbour were dragged down with it. The most famous natural disaster of classical antiquity was also caused by the region's seismic activity: the eruption of Mount Vesuvius on 24th August 79 AD. Earthquakes had been growing in strength and frequency for ten years before the catastrophic eruption, which started with a colossal bang as the magma and pressure that had been building for thousands of years finally burst through the crater of Vesuvius. This was just the beginning, and hours later an even bigger explosion blew off the entire cone of the volcano, throwing a massive mushroom cloud of rock twenty-seven miles into the sky. The power of the explosion has been calculated at 100,000 times greater than the Hiroshima nuclear bomb.

The fury of the volcano fell most strongly on the ancient Roman city of Pompeii – a large and important trade centre located in the Bay of Naples – that disappeared completely, until it was unearthed in 1755, a perfectly preserved Greco-Roman city frozen in time. Buildings were razed to the ground, the population was crushed or suffocated, and the entire city was buried beneath a blanket of ash and fiery pumice. A first-hand account of the eruption has survived in the form of two letters by Pliny the Younger, who witnessed the event from his uncle's villa at Misenum:

You could hear the shrieks of women, the wailing of infants and the shouting of men; some were calling their parents, others their children or their wives, trying to recognize them by their voices. People bewailed their own fate or that of their relatives, and there were some who prayed for death in their terror of dying. Many besought the aid of the gods, but still more imagined there were no gods left, and that the universe was plunged into eternal darkness for evermore.[2]

Buffeted by the Gods

Life in the ancient world was inherently risky, but what were people's attitudes to risk? Early philosophers wrote extensively about the nature of fate, providence, chance and free will. While languishing in prison awaiting execution and suffering from depression, the Roman senator and philosopher Boethius wrote extensively about these themes in his famous treatise, *The Consolation of Philosophy*. Described by one translator as 'one of the most popular and influential books in Western Europe from the time it was written, in 524, until the end of the Renaissance', it introduced ideas that run through the philosophy of late antiquity. In it, he suggests that the world is governed by a divine plan which guides the direction of events, that fate and chance operate on a day-to-day level within that big picture, and that humans have the freedom to choose but only within the boundaries of these forces. He imagines a non-interventionist God acting as 'a helm and rudder, by which the fabric of the world is kept stable and without decay'. This was a sharp contrast to Stoicism, whose followers believed there was no point in worrying about things because everything is predetermined. Boethius's philosophy offered some hope that we *can* influence events or our own fortune by exercising our will towards achieving good or happiness. It also allows room for chance, the happening of events without apparent cause, to exist. Unexpected occurrences happen all the time, but according to 'the order proceeding from the connection among all things. This order emanates from its source, which is Providence.'

But what did the average citizen in ancient Rome or Greece worry about? How different were their anxieties and fears from the things that preoccupy us today? And how was risk really viewed, beyond just a basic understanding of common hazards or dangers and how to manage them? Throughout the ancient world, a deeply held belief in the gods, ritual and mythology was shared by everyone from slaves to aristocrats. They believed that the gods frequently interfered in people's lives, were often mischievous and sometimes helpful. Today, in central Athens, if you walk away from the tourist hoards, cafés and gift shops surrounding the monuments of the ancient Agora, you will find a quiet back street that takes you past the Gate of Athena into a calm, peaceful oasis next to the Museum of Greek Folk Musical Instruments. It's the perfect place to sit on one of the benches and escape the intensity of the city for a few minutes. This small square of streets overlooks an extraordinary monument: the Tower of the Winds, also known as the Horologion of Andronikos Cyrrhestes after the Macedonian astronomer who designed it. A beautiful octagonal tower made from Pentelic marble, it was an early type of clock tower and weather station. It originally contained sundials and a water clock and was topped by a giant wind vane. What makes it so striking today is the well-preserved allegorical frieze that runs along the eight sides and depicts the wind gods, or Anemoi, one for each of the eight points of the Athenian compass. On one side, you can see a beardless young man pouring water from his hydria (this is Notus, the bearer of rain) and a stern old man wrapped in a cloak (Eurus, the bringer of storms). On the other side, you can see an old man blowing through a conch shell, representing Boreas, the god of the north wind and the harshest of the Anemoi. Walking past this high-tech meteorological station every day would have been a constant reminder to the people of Athens of the power of these fickle deities.

It was believed that the Anemoi governed the seasons and weather conditions and are most often invoked when it comes to the sea and sailors, who are more subject than most to the vagaries of weather. In Homer's *Iliad*, Zephyrus, the west wind, brings gusts and gales that churn the ocean:

As crosswinds chop the sea where the fish swarm, the North Wind
and the West Wind blasting out of Thrace in sudden, lightning

attack, wave on blacker wave, cresting, heaving in a tangled mass
of seaweed out along the surf.[3]

In the *Odyssey*, Homer's hero Odysseus struggles against the
machinations of the gods who decide what fate befalls him and
whether he will ever succeed in making the long journey home to
Ithaca. He is lost at sea, buffeted by the winds, thwarted by gods
and sometimes punished for his actions. But, in the end, Zeus takes
pity on poor Odysseus and sends a storm to bring him safely home
after his long voyage. The state of being constantly buffeted by the
gods with no control over how life unfolds was a common theme
for the main characters in Greek drama of that period and likely
represents how many ordinary people felt at the time.

To mitigate the uncertainty and helplessness that might accom-
pany this belief, the ancient Greeks turned to oracles for help. The
oracles were women chosen by the gods who could give divine
insight, wise counsel or make prophecies about the future. They
ranged from the wise woman or fortune-teller that could be found
in the marketplaces of most towns to the great oracles people
would travel hundreds of miles to consult. The most famous, the
Oracle of Delphi (the Pythia), was thought to be the mouthpiece of
Apollo, the god of prophecies, and was a position occupied by the
high priestesses of a temple dedicated to him at Delphi. From the
seventh century BC to the fourth century AD, the Delphic Oracle
was considered the most prestigious and influential amongst the
Greeks and famously foresaw the death of King Leonidas, 'torn
limb from limb' by Persian arrows at Thermopylae.

While kings and emperors consulted the oracles to help plan
their war strategies or decide what to build, ordinary people also
sought their advice and would wait in long queues for hours to ask
questions. They provided answers in an uncertain and sometimes
dangerous world, comforting people who felt they could not influence their destiny. If the gods control your fate, who better to ask
what you should do when disaster strikes or you have a problem?
Research by Professor Mika Kajava from the University of Helsinki

has found that the most common questions the ancient Greeks asked the oracles are not so different from the ones modern Greeks have. Questions about happiness in your marriage, whether you will have children and whether you would find a good job were among the most frequently asked. People also enquired about the safety of long journeys and how to ensure continued good health. The ancient Greeks often asked the oracles: 'To which god should I pray to see my business prosper?' While the answers might give people hope and help alleviate anxiety, they were almost always enigmatic or ambiguous. The oracles never got it wrong (and made sure of that). As a way of hedging their bets, the oracles gave vague answers that were difficult to interpret, claiming that there is inherent ambiguity in the god's signs and limits to a mortal's access to information about the future. If the advice you got didn't help or made things worse, it isn't because the oracle got it wrong but rather that you misinterpreted the message in the first place.

The Die Is Cast

Gambling is one of mankind's oldest activities, and the attitude of the average citizen towards gambling in different civilizations throughout history can tell us a great deal about how they conceived risk and uncertainty. It is well known that gambling is more prevalent in societies with widespread belief in gods and spirits whose benevolence might protect them from the hazards associated with games of chance. The word 'hazard', familiar from everyday usage but also an essential technical term in professional risk analysis, comes from the Arabic word 'al zahr' meaning dice. The literal meaning is to put something at stake in a game of chance. The 'd' was added later in the French, and the sense evolved from 'chances in gambling' to 'chances in life'. It was first recorded in English in the 1540s, meaning 'chance of loss or harm, risk'. This simple word shows how humans associated danger with luck or uncertainty. Dice and their forerunners are the oldest gaming implements known and were used throughout the ancient world for gambling.

Despite the claim by Sophocles that dice were invented by the Greek prince Palamedes during the siege of Troy, they existed long before recorded history and their exact origins are murky. Early forms of dice can be seen in Egyptian hieroglyphics; a fourteen-sided die was found in a 2,300-year-old tomb in China, and dice games were written about in the Sanskrit epic, the *Mahabharata*, composed during the third century BC. They evolved from knucklebones or astragals, the anklebones of sheep, buffalo or other animals, sometimes painted on four sides, and initially used by primitive people for the casting of lots to divine the future. Knucklebones were also used to play games and for gambling, although they eventually gave way to the popularity of dice.

The oldest known surviving dice was unearthed during an excavation at the Burnt City, Shahr-e Sukhteh, one of the richest archaeological sites in Iran believed to have been the capital of an ancient civilization that flourished on the banks of the Helmand River around 4000 BC. Over the past one hundred years, excavations across the fifty-five hilltops of the site have produced an estimated four billion artefacts, including some remarkable discoveries, such as the world's first artificial eye and a goblet painted with what experts believe is the earliest known animation. In 2004, the team found a backgammon board made of ebony and engraved with a snake coiled in a pattern that made twenty slots for the game. Next to it was a clay pot containing sixty turquoise and agate markers and a pair of six-sided dice.

By Greek and Roman times, gambling with dice or other games was a favourite form of entertainment enjoyed by patricians and commoners alike. It was played in taverns that had special rooms for the purpose (a predecessor of the casino), open-air in public spaces, in private homes and even in army barracks. According to Pausanias's *Description of Greece*, published in the second century AD, there was a Temple of Fortune dedicated to dice in the city of Corinth. Gambling in Rome became so rampant that it was eventually prohibited, being seen as a serious threat to the moral and social well-being of the city-state. Worries about its addictive properties and the effect it could have on people's lives were championed by its most vocal critics, but despite this (or perhaps because of it), it continued to thrive. The Romans embraced the inherent risk involved, the attendant rush of adrenaline it gave and the emotions that surrounded the exciting vision of winning versus the fear of losing.

The Cambridge historian Mary Beard believes that the Romans operated a model of risk explicitly derived from their understanding and relationship with dice games. According to Beard, they lived in an aleatory society (from '*alea*', meaning bone or dice), one that views risk as they do the throwing of a dice. They actively used the image of dice to help them understand the world around them and deal with events as they come along, 'facing' and 'parading' the luck of the dice in everyday life. Nowhere is this better shown than in Caesar's crossing of the Rubicon, the river that marked the border between Gaul and north-east Italy. This meant passing the point of no return and initiating civil war. He led his troops across on 10th January 49 BC, famously saying, '*Alea iacta est*' ('the die is cast'). His expression has been interpreted to mean that he had made a move and there was no turning back – but he wasn't abrogating responsibility and simply hoping for the best. Semantic studies have shown that the words used metaphorically mean 'an act of risking or state of risk, chance, hazard, gamble, uncertainty; something which involves uncertainty, a risky enterprise or purchase'. Far from accepting their lot as mere dupes or playthings of the gods, the Romans had weaponized their understanding of the dice to develop an attitude that manages risk or danger by facing it head-on. They didn't try to measure, analyse or mitigate it. Nor did they simply accept the will of the gods or employ an oracle to avoid making decisions. Caesar's statement deliberately paraded the uncertainty of their endeavour, looked danger square in the eye and celebrated it as part of human existence. In doing so, Caesar was not only risking his own future but the fate of the Roman Republic.

Out of Arabia

The Arabian Desert is a vast, seemingly limitless expanse. The fifth-largest desert in the world, it occupies most of the Arabian Peninsula, stretching from Jordan and Iraq in the north to Yemen and the Persian Gulf in the south. At its heart lies the Rub' al Khali, or Empty Quarter, one of the world's largest continuous bodies of sand. Until around 300 AD, the lost cities of the Empty Quarter (now buried under miles of sand) relied on the trade of frankincense before desertification made even camel crossings all but impossible. Despite its deprivations, the Arabian Desert has been inhabited since the Pleistocene, around 2.6 million years ago, and for the past 3,000 years by various tribes collectively described as Bedouin, from the Arabic *Badawi* meaning 'desert dweller'. They inhabit a strange world where the air is dry and dusty; the bright, clear skies are dominated by the sweltering sun during the day and filled with countless stars at night. The landscape is composed of great rocky plains or endless miles of dunes interspersed with occasional mountain ranges or black lava flows. Far from being daunted by the scale of the desert and its hostile conditions, the Arab people thrived – breeding camels and cultivating some of the harshest terrain imaginable to grow crops.

Arabia is inextricably linked with the dramatic rise of Islam in the seventh century. Mecca, situated on a wide sandy valley, was a desert town of trade and pilgrimage with a vital well at its centre; Medina was a date palm oasis, and the prophet Muhammad was a caravan driver of Bedouin training. The pages of the Qur'an show how his mind was filled with images of the desert: the mirage being a manifestation of the 'false hopes of unbelievers', sandstorms an emblem of destruction and the oasis a symbol of paradise.[4] The rare desert oases were indeed a paradise for the Bedouin, being the only source of water. If the Bedouins stayed at one oasis too long, they would exhaust the water supply. So, they had to adopt a nomadic way of life, moving from one oasis to another and travelling long distances across the desert. The humble camel enabled their way of life, being more important to the Bedouin than gold. Well adapted to desert life, it needs little water to survive, can carry heavy loads over long distances, and even provides nutrition in the form of milk and meat.

Their expertise with the camel and unique ability to traverse the desert helped them establish a vast trading network connecting East and West. The growing prosperity of the southern part of the Arabian Peninsula in the second century led the Romans to call it *Arabia Felix*, meaning 'Happy' or 'Flourishing' Arabia, mainly because it was rich in incense, myrrh and saffron – all of which were highly sought after at the time. Arab traders brought these products to Roman markets by the camel caravans of the Incense Route, which was said to transport 3,000 tonnes of incense every year and, according to Pliny the Elder, took sixty-two days to complete. Similar trade routes crisscrossed the Arabian and Saharan Deserts, linking Africa, India and the Far East with the Greco-Roman world of the Mediterranean and contributing to a burgeoning trade in spices, gold, ivory, precious stones and textiles. For hundreds of years, this overland network formed the most important trade route in the world, providing an important source of tax revenue to the Roman treasury, and by the twelfth century, some of the largest caravans were said to contain 12,000 camels.

In addition to the importance of the camel, being able to endure incredibly long journeys through the desert depended on the Arabs' desert survival skills: wearing long black robes that absorb heat and keep it away from the body, hoods to protect the face in sandstorms, the ability to find water and to make it last, travelling at night and resting during the hottest part of the day. They were also excellent navigators: experts at reading the wind and the stars. Instead of a compass, Bedouins used sand dunes, which form perpendicular to the prevailing wind direction, to point the way. They understood the direction of the rising and setting sun, using the shadow it cast as a makeshift sundial and their knowledge of the stars to guide them safely to the next oasis.

And it is He who placed for you the stars that you may be guided by them through the darknesses of the land and sea. We have detailed the signs for a people who know.

– Qur'an: 6:97

Perhaps because of the need to navigate the deserts, ancient Arabic astronomy was the most advanced in the world. Building on the work of the early Greek and Indian pioneers, they developed new ways of measuring the position of the heavenly bodies, improved on the calculations in Ptolemy's *Almagest* and created new mathematical models to better understand the movements of the sun, moon, stars and planets. From the eighth century to the thirteenth century, as Europe struggled through the Dark Ages with little cultural advancement, the Golden Age of Islamic science flourished, making great advances in astronomy, mathematics, medicine, geography and cartography. Celestial navigation depends on accurate star charts that help the navigator identify particular stars within constellations. The *Book of the Fixed Stars*, compiled by Persian astronomer Al-Sufi around 964, was one of the first to synthesize the comprehensive star catalogue in Ptolemy's *Almagest* with the Arabs' traditional indigenous knowledge of the constellations and new learning brought by the desert trade routes and the Silk Road. Of the 300 stars that have been named, over 200 originate from Arabic names and have Arabic pronunciation: Aldebaran, Algol, Deneb, Vega and so on.

Albrecht Dürer's *The Northern Hemisphere of the Celestial Globe* (1515).
Woodcut print depicting the constellations of the northern hemisphere. In
each corner, the four authorities, on whom the constellations are based, are
pictured: Aratus Cilix (Aratus of Soli); Ptolemaeus Aegyptus (Ptolemy);
M Mamlius Romanus (Marcus Manilius); and Azophi Arabus (Al-Sufi)

The earliest Arabic poetry from around 1,500 years ago survives
through the Bedouin oral tradition and gives us an insight into what
the night sky was like at the time. There was no North Star to guide
you; instead, a trio of stars danced around the North Celestial Pole,
marking the direction North. Polaris, or the North Star, was one of
these and was called the 'Goat-Kid' (Al-Jady), and the others were
a pair of stars called 'The Two Wild Cow Calves' (Al-Farqadan).
The Islamic scholar Ibn Qutaybah Al-Dinawari, who died in the
year 276, wrote *Settings in Arab Seasons*, a book which covers the

stars and their orbits and describes the importance of stellar navigation in the desert:

> A nomad man accompanied me in the desert one night; I asked him about the places of an Arab tribe and their waters, and he showed me every place with a star and every lit place with a star. He sometimes referred to a star and called it its name; at other times he said to me, 'You see it.' At yet other times he told me, 'Follow the star so and so;' i.e., walk with the star so and so until you reach them. I saw stars leading them to their needs, as clear paths lead people to buildings.

The necessity for water and the powerful economic desire to trade drove the Arabs to conquer their natural fear of the desert and to become great explorers and navigators capable of travelling vast distances into the unknown, combating the extremities of climate, weather and circumstance. It was a dangerous way of life, and they put their fate in the hands of God. This is where the first notion of risk emerged, as a religious concept that helped the Arabs cope with the uncertainties of the future and put their trust in something more powerful and reassuring than the advice of oracles or accepting that life is just a game of chance. The Arabic word 'risq' is the earliest form we can trace of the modern word 'risk' and is probably the root source of all later variations. It is part of the story of the Islamic faith, which emerged from the deserts of Arabia and appears in the Qur'an and oral traditions of the region. It translates as 'provision' or 'gift' from God, and more broadly means: 'what can be provided by the Providence which can be good or bad for the Orthodox Muslim'.

The poet Ibn Arabi (1165–1240) associates risq with 'happy trust', indicating that it confers a sense of happiness and contentment, while the Qur'an puts it in the context of travel by land and sea. There is no need to worry about the outcome of a perilous journey when you have the 'risq of God' and its spiritual blessing on your side. The Qur'an makes it clear that risq assists with material things: eating, drinking, trading, marriage and travel. Indeed, as Islam expanded and conquered more lands after the seventh century, it came to incorporate a commercial and economic meaning – indeed, in Al-Basra, a city founded in 636 as a garrison for Arab armies, the central market was called 'The House of Risq'. Perhaps these were the biggest worries of early Islamic communities. Before starting a long journey, the Bedouins and the great camel caravans of the Arabian Desert could put their trust in God and receive the *risq* (or blessing) for themselves and the goods they transported. Alleviating their fears, it was a kind of spiritual underwriting or insurance.

Many historians have said that the Arabs were a desert people afraid to travel by sea, but as the Islamic world expanded into the Mediterranean and beyond, maritime voyages became more common, and Arabs swapped the boundless expanses of the desert for the dark, broad seas. It is true that the first caliphs initially feared the sea, but their expertise at navigation and growing appreciation of *risq* led them to conquer the oceans as they did the deserts, fast becoming a major power in the Mediterranean and Indian Ocean. Overtaking the Byzantines, the Mediterranean became an Arab sea from the ninth to the twelfth century. Their belief in *risq* – that God would guarantee the future no matter how difficult or uncertain the journey – became an important tool to manage the human and commercial elements of maritime trade.

The Age of Discovery

Evidence of maritime trade can be found as far back as the Stone Age when different populations around the world independently developed dugout canoes for fishing or to trade over short distances. Pacific islanders developed outriggers, crab-claw sails appeared in Southeast Asia sometime before 2000 BC, early dhows carried goods across the Indian Ocean, and the ancient Egyptians built some of the earliest warships. Around 5,000 years ago, the first trade routes were established, starting with safer coastal routes then crossing larger expanses such as the Arabian Sea or the Bay of Bengal to connect prospering trading centres. These early sea lanes were an important alternative to the camel caravans that took a long time and were often attacked by bandits. Around the same time, the Romans were building fleets of ships that could cross the entire Mediterranean in less than a month, making it much cheaper to transport goods such as grain and construction materials. The expansion of the Arab world, initiated by Muhammad in the seventh century, created new trade routes through Europe, Africa and Asia. By the fourteenth century, maritime trade in the Mediterranean was flourishing, with the Venetian Republic and the city of Genoa emerging as the dominant powers.

As this new industry grew, so too did the legal and financial structures needed to conduct business between Arab and European ports. There was a busy exchange of ideas and words between these two cultures, and the modern concept of risk first entered European thinking through this maritime route. The Arabic phrase *risq* became the Latin *resicum* and first occurred in a Genoese maritime contract dated 1248. In the contract, it was used to describe who was responsible for any damage that might occur to the ship or its cargo. In this and later similar sea contracts, the term developed to refer both to the objects *at* risk, such as a ship, and the objects *of* risk. This included piracy, wreck, theft, violence and 'any other occurrence of danger'.[5] The earliest contracts were primarily concerned with who would guarantee any damage or loss – be it the owner, customer or the ship's captain – and soon led to the very first maritime insurance policy we know of, for a ship called the *Santa Clara* sailing from Genoa to Majorca in 1347. In it, Georgius Lecavellum claims to have received from ship owner Bartholomeus Bassus 107 pounds of silver. In return, he makes a commitment to undertake all the risks of the navigation in the following way:

> I assume personally all risk and responsibility for the stipulated amount until the boat reaches Mallorca ... and promise to make such compensation and also promise to pay you and incur a penalty of double the amount previously provided in addition to the restitution of damages and expenses that may arise...

So, if the ship safely reaches Majorca, the ship owner will receive a refund of his money within six months. Maritime insurance was born with a commercial activity that specialized in the exchange of risk with a specific monetary value. The idea of risk had evolved from a religious concept to one associated with the technicalities of trade and became firmly embedded in the foundations of the shipping industry. By the fifteenth century, risk was a term in widespread use by anyone involved in navigating the seas.

The Reconquista, the long and bloody campaign to liberate parts of Spain and Portugal from Muslim invaders, was almost over by 1400 and finally ended after the surrender of the last Muslim stronghold, Granada, in 1492. The Kingdom of Portugal had doubled in size, but the war had been expensive, and the royal treasuries were depleted, so new sources of income were desperately needed. At the same time, trade between Europe and the East was a vast enterprise, bringing enormous wealth and prosperity to the European kingdoms, especially the maritime superpowers Venice and Genoa. The main land route, the Silk Road, was closed to Christian traders following the fall of Constantinople to the Ottomans in 1453, leaving sea routes the only viable way to access eastern wares and spices. Henry the Navigator, the fourth child of the Portuguese King John I, was the driving force behind Portugal's rapid maritime expansion – exploring the North African coast, pushing further and further south, and towards India to capture a share of the spice trade. The royal family were quick to offer funding and patronage to dangerous voyages that might lead to the acquisition of new lands and gold while also converting new groups to the Christian faith. As a young man, Henry spent time exploring North Africa and became obsessed with finding the source of West African gold and trying to wipe out the Barbary pirates that frequently raided the Portuguese coast. He was fascinated by cartography and oversaw advances in navigation that enabled his seafarers to reliably harness the trade winds of the Atlantic, ushering in the Age of Discovery.

This was the environment into which Christopher Columbus was born in 1451 in Genoa, the son of a wool merchant. Surrounded by the sights and sounds of the maritime city, he grew up with a passion for sailing and navigation and joined the merchant navy as a teenager. As a young sailor, he travelled around the Mediterranean to Africa and Greece, learning the ropes and honing his skills. His time as a sailor came to a premature end when his ship was attacked and burned by pirates just off the coast of Portugal in 1476. Clinging to a plank of wood, Columbus was able to swim to shore and made his way to Lisbon, where he eventually settled. In Lisbon, he studied cartography, mathematics, astronomy and navigation, quickly becoming an expert in these emerging fields. The combination of this knowledge and his deep understanding of sailing made him unique. Michele de Cuneo, who sailed with him, said:

> By a simple look at the night sky, he would know what route to follow or what weather to expect; he took the helm, and once the storm was over, he would hoist the sails, while the others were asleep.

While others sought new routes to India and Asia by sailing east and attempting to round the Cape of Good Hope on the southern tip of Africa, Columbus looked to the west. Many historians have speculated on what gave Columbus the inspiration, will and courage to go where no man had gone before. Some believe he had secret information provided by an 'anonymous' pilot who had been swept away by a storm and ended up on the coast of Brazil. That the Earth was a sphere had been known since antiquity. In the third century BC, Eratosthenes was the first to accurately calculate the circumference of the Earth (which he put at 252,000 stadia, equating to roughly 24,000 miles). But Columbus had done his own calculations and believed it was much less and that you could reach Asia faster by crossing the Atlantic rather than taking the Cape route. In 1484, he set out a proposal for a pioneering voyage west to reach Asia via the Canary Islands, making use of the trade winds and the latest nautical charts. It took years to find support for the expedition, with experts criticising his estimate of the nautical distance, which they believed to be woefully short of the mark. Eventually, he reached an agreement with the Spanish crown in April 1492.

Illustrative woodcut from the Latin edition of Columbus's letter announcing his discovery printed in Basel in 1494.

Like modern business entrepreneurs, Columbus was partly motivated by the possibility of success, great wealth and personal fame. His patrons, King Ferdinand and Queen Isabella, promised that, if he succeeded, he would be rewarded with the title of 'Admiral of the Ocean Sea', appointed viceroy and governor of all the lands he claimed for Spain and entitled to ten per cent of all the revenue from them. The anthropologist Carol Delaney has argued that he was a religious extremist whose actions were governed by a desire to convert all races to Christianity, conquer the Holy Lands and prepare for the coming apocalypse. Like the Arab caravan traders, Columbus's approach to risk was largely based on his faith, belief in providence, and unique navigation skills. He pushed the maritime definition of risk beyond the realm of insurance and guarantees towards it being seen as something bold, courageous, adventurous and even entrepreneurial. Having been near or on a ship ever since he was a young boy, he understood the enormous risks involved in perilous voyages of exploration, yet was always optimistic about the potential outcome, knowing that God would ensure good fortune.

Columbus embarked on his first Atlantic voyage on 3rd August 1492 with three ships, the *Santa Maria*, the *Pinta* and the *Niña*. After restocking in the Canary Islands, they sailed across the ocean for five weeks before finally sighting land on 12th October. The island they discovered was named San Salvador and was the most easterly of the islands that make up the Bahamas, just 100 miles off the coast of Florida. They had found the New World, although Columbus believed it was the 'Indies' and that they had reached Asia. He held firm to this belief through all his subsequent voyages. 'The Admiral was the first to open the gates of that ocean which had been closed for so many thousands of years before. He it was who gave the light by which all others might see how to discover,' wrote historian Bartolomé de las Casas half a century later in a comprehensive account of the voyages.

The Spanish version of risk, *riesgo*, appears many times in Columbus's letters and sea contracts. Like other sea contracts of the period, it is considered a probabilistic relationship between the things at risk (the ship and the cargo) and the hazards they face. In this case, the King and Queen would cover any damage caused by big storms or losses due to pirates. Risk helped the Arabs cross the vast spaces of the desert and was now helping explore the oceans and find new worlds. It had passed from the Arabian Peninsula through the Mediterranean and now set foot in America. It had transformed from a purely religious concept to being part of the language of maritime law and commerce. According to Gaspar Mairal, it was no longer a relationship with God but 'a relationship between an owner and his merchandise, or between an insurer and the insured object. Finally, the trust in God was substituted by a contract or an insurance policy.'[6] Through Columbus, risk was directly associated with something ground-breaking, adventurous and entrepreneurial. The next phase in the evolution of risk was brought about by progress in the mathematics of probability.

The Man Who Broke the Bank at Monte Carlo

As the world holds its breath in the late summer of 1939 following Germany's belligerent occupation of Czechoslovakia, Winston Churchill is sitting in the casino room at the Hôtel de Paris Monte-Carlo in the exclusive principality of Monaco. The atmosphere in the grand hall is electric. Shrouded in a fog of cigar smoke, he shouts, 'Nothing is going right!' as he listens to the roulette wheel's spinning clicks and the bouncing ball preparing to land on some random number. 'You must stop,' suggests the casino director, but he keeps playing and keeps losing, in thrall to the obsessive magic of the game. 'I'll pay my debts to you tomorrow,' Churchill replies as his losses mount.

The next day, he was gone. He raced back to London as Britain prepared for the outbreak of war. It was just as well because he was broke and couldn't have paid the 1.3-million-franc debt – he was overdrawn at the bank, owed interest payments on his loans, was late with his taxes and was in hock to everyone from his shirt-maker to Fortnum and Mason, who supplied him with Dundee cake. Despite the state of his bank account, he loved Monte Carlo and, like the ancient Romans, was addicted to the highs and lows of games of chance. Long after the war ended, at the age of seventy-one, Churchill remembered his debt to the casino and wrote them a cheque for 1.3 million francs. In tribute to the Hero of Europe, the Société des Bains de Mer has never cashed the cheque.

Such was the glamour of the most famous casino in the world. Opened in 1865, the spectacular building was later expanded to include the Opéra de Monte-Carlo, designed by Charles Garnier, the architect responsible for the iconic opera house in Paris. In the Belle Époque style that typified a period of cultural and artistic optimism, the impressive façade is complemented by intricate stained-glass windows, ornate sculptures and dramatic chandeliers. The height of elegance and sophistication, it attracted the beau monde including the Prince of Wales (Edward VII), Alexandre Dumas, Prince Napoleon, Aristotle Onassis (who used to lend Churchill chips to play with) and, of course, James Bond.

The luckiest man ever to visit the casino was Charles Wells, a British inventor and entrepreneur. On the afternoon of 28th July 1891, all eyes were on this short, ordinary-looking man with a bald head and dark moustache, who was slowly being hidden behind the huge pile of banknotes he was accumulating. Having arrived just hours before, he was enjoying one of the most remarkable winning streaks ever seen, and the roulette table kept spitting out his numbers. Crowds gathered to watch as he stayed at the same table all day, never pausing for food or drink until the casino closed at 11 p.m. During one roulette session, he won twenty-three out of thirty consecutive spins and, during the visit, amassed a fortune worth around £4 million in today's money. Over five days, he broke the bank multiple times, forcing tables to be closed because they did not have enough money to pay his winnings.

The story of his success travelled far and wide, with Wells being hailed as a kind of people's champion, a symbol of hope that anyone could win a great fortune if they were lucky enough. He was immortalized in a hit music-hall song by Fred Gilbert, 'The Man Who Broke the Bank at Monte Carlo':

I to Monte Carlo went, just to raise the winter's rent.
Dame fortune smiled upon me, as she'd never done before;
And I've now such lots of money I'm a gent.

Just a few years later, in 1893, Wells was in Wormwood Scrubs prison in London, convicted of fraud for convincing people to invest in fictional inventions he never made. Still loved by the public, a big crowd gathered outside the prison windows to sing the Gilbert song to him. Wells later claimed it was all down to luck, but many have suspected that he colluded with the management or had some kind of gambling system that gave him a unique understanding of the odds. Such a system has always been the holy grail for gamblers.

During the Renaissance, gamblers would discuss the odds of something happening, for example, 'six to one', and maritime insurance premiums were based on a subjective estimate of risk, but no one had yet developed a way to calculate them. The Italian polymath and gambler Gerolamo Cardano was the first to investigate games of chance and, in doing so, devised the foundations for the mathematics of probability. Cardano was a renowned physician, physicist, chemist, astronomer, philosopher, writer and gambler who had a huge influence on sixteenth-century science and mathematics. Born in Pavia in 1501, his father, Fazio Cardano, was primarily a lawyer but also a gifted mathematician who studied perspective and was a close friend of Leonardo da Vinci. Cardano was clearly something of a child prodigy and developed a reputation for being hot-tempered and single-minded as well as highly intelligent. In true Renaissance style, he applied his intellect to many different areas of science and philosophy and wrote over 200 learned works during his lifetime. He was best known for his medical research and became one of the most sought-after doctors in Milan. In 1552, he travelled to Scotland to treat the Archbishop of St Andrews, who had been suffering from a mysterious disease for almost ten years that left him short of breath and unable to speak. Realising it was a type of asthma likely brought on by various allergies, Cardano treated him successfully. The archbishop was so grateful he paid him 1,400 gold crowns, and his fame spread – resulting in frequent demands from Europe's nobility, including the kings of Denmark and France and the queen of Scotland (all of whom he turned down, as he didn't want to leave Milan).

Cardano was always short of money and turned to gambling while at university to supplement his income. His knowledge of probability meant he was systematically able to win more than he lost, generally turning a regular profit. But gambling as a young man could be dangerous and he always carried a knife, once using it to slash the face of an opponent who he believed was cheating. What started as a vital source of income soon became an addiction that lasted his entire life. In his autobiography, he confessed to an 'immoderate devotion to table games and dice… During many years… I have played not off and on but, as I am ashamed to say, every day.'[7] While it sometimes brough him great losses, it was also the inspiration for one of his greatest works, *Liber de ludo aleae* (*Book on Games of Chance*), written in 1565. In it he explains that you can reliably calculate the odds of dice throws as the ratio of favourable to unfavourable outcomes, implying that the probability of an event is given by the ratio of favourable outcomes to the total number of possible outcomes. Cardano was the first to establish a mathematical framework for looking at risk and uncertainty, producing something that could be used (as he no doubt did) as a risk-management tool for gamblers as well as a guide to the rules of probability. It took another child prodigy, Blaise Pascal, together with Pierre de Fermat and the Chevalier de Méré, to devise a theory of probability with hard numbers in 1654. Again, it was created to solve a problem in a game of dice: if you throw a pair of dice twenty-four times, how profitable is it to bet on a double six appearing? The theory states that if we suppose a game has 'n' equally probable outcomes, of which 'm' outcomes correspond to winning, the probability of winning is 'n/m'.

Throughout history, risk has been used as a way of visualising possible futures so that people can make the best decisions possible and manage their behaviour. Now, hard numbers could be applied to forecasting the future, paving the way for the fields of risk analysis and risk management. As the historian John Ross writes, 'Probability theory and the discoveries following it changed the way we regard uncertainty, risk, decision-making, and an individual's and society's ability to influence the course of future events.'[8] God had been replaced by mathematics.

The Risk Society

Quantified risk analysis is now a common tool used in almost all areas of our life. In healthcare, it helps people to better understand the potential harm of certain behaviours and the benefits of life-saving drugs; it helps keep us safe at work or when using transport, can predict hurricanes so we can protect our homes better, and is used to stop us making huge losses when we make financial investments. We live in a world that uses probability and analysis to understand the risks we face and to manage them. But that only works for the risks we know well and have good information about. What about new or unforeseen risks? In 1921, Frank Knight made an important distinction between risk and uncertainty. For Knight, risk applies to situations where we don't know the outcome but have enough information to accurately calculate the odds, whereas uncertainty is the lack of any quantifiable knowledge. 'Uncertainty must be taken in a sense radically distinct from the familiar notion of risk,' said Knight, 'from which it has never been properly separated... It will appear that a measurable uncertainty, or 'risk' proper... is so far different from an unmeasurable one that it is not in effect an uncertainty at all.'

Acid rain, the Chernobyl explosion and the surprising discovery that greenhouse gas emissions had caused a hole in the ozone layer over the Antarctic were the defining environmental issues of the 1980s. Concerns over the impact of pollution on human health and the devastating effect of human behaviour on the planet helped environmentalism to grow into a global political force. It led sociologists Anthony Giddens and Ulrich Beck to develop the concept of the *Risk Society*. They described a fundamental shift needed to cope with the challenges and dangers of modern society. Whereas in the past, most of the biggest risks facing humanity had been natural disasters or accidents, in the modern world, they are being manufactured by human action. According to Giddens, the Risk Society is: 'a society increasingly preoccupied with the future (and also with safety), which generates the notion of risk'.[9]

Ulrich Beck, a professor at the University of Munich, first published his ideas in German in 1986, immediately after the Chernobyl disaster. His book, *Risikogesellschaft*, was later published in English as *The Risk Society: Towards a New Modernity* in 1992. He argues that society is increasingly occupied with managing the risks it has itself produced. The danger is not direct or intentional, in the way war or terrorist attack is, but unforeseen and unknown. Risk is not about catastrophe, but about anticipating the future through imagination and visualization, creating hypotheses about possible futures and using that understanding to direct our responses. Unlike Knightian economics, this approach advocates we incorporate the unknown and unforeseen into our model of risk. To do this, we must develop our imaginations. He believed that large-scale risk has no borders: it cuts through nationalities, cultures and religions and demands cooperation and a cosmopolitan approach to protect the interests of the citizens of all nations. In a *Risk Society*, national states and civil society organizations are empowered to take action to keep us safe and to harness the opportunities in the future we imagine.

The WEF *Global Risk Report* reminds us that we live in a world that is more interconnected and interdependent than ever before, with today's crises, such as those relating to the cost of living and energy supplies, compounding the economic aftershocks of the Covid-19 pandemic and the climate emergency. Echoing the main principles in *The Risk Society*, it highlights the need to overcome geopolitical tension and to cooperate across borders as well as the importance of finding better ways to predict or visualize the future so we can build a more resilient world.

Chapter 3

The Ancestor Stones

That men do not learn very much from the lessons
of history is the most important of all the lessons
that history has to teach.

– Aldous Huxley, *Collected Essays*

From mid-October to November every year, Japanese families
head to the countryside. They go for the changing autumn colours,
known locally as *koyo*, when trees transform into stunning shades
of yellow, gold and red. *Koyo*-seekers flock to Fukushima, Japan's
third largest prefecture, with its mountains, rivers and shrines. Sites
like the Byobuiwa Crags are popular places for a picnic or a hike.
A rock formation in the southern area of Fukushima that has been
eroded over millions of years, the crags are made of light-coloured
stone that contrasts sharply with the azure green and blues of the Ina
River that gushes alongside. In autumn, the trees that perch above the
cliffs create a riot of colours as a backdrop for walking and taking
pictures. Leaf-peepers also head to Iwaki City, where you can find
the Buddhist Shiramizu Amidado Temple, built in 1160 and officially
designated a national treasure. Surrounded by Japanese maple trees
that turn a vibrant crimson red, it sits in an ancient paradise garden
and has been described as 'one of the most tranquil places in the
world'. It is hard to believe that such a beautiful and peaceful area
was the focus of Japan's worst catastrophe in recent history – its
name now synonymous with nuclear disaster.

Fukushima is an agricultural district located on Japan's main
island in the Tohoku region, just 250 km north of Tokyo. It is an

area of rice fields, forests, streams and mountains with a coast-line famous for fishing and seaweed production. A leading centre of silk production for hundreds of years, it flourished in the nineteenth century as Japan industrialized, using the newly built Tohoku-Honsen railway to transport large amounts of raw silk to coastal areas where it was exported. It has long been central to Japan's energy production – at one time producing most of the coal consumed by Tokyo and later being home to one of Japan's first hydroelectric power stations.

In the immediate aftermath of the Second World War, the dev-astated Japanese economy rose quickly from the ashes, catching up with the USA and Great Britain, two of the most advanced industrial economies in the world. Rapidly adopting new technolo-gies from the West, Japan's leaders transformed its manufacturing industry and understood the importance of a pipeline of skilled workers, many of whom were retrained agricultural labourers. The 'Japanese Miracle' made it the fastest-growing economy after the war, soon becoming the second-largest in the world.

To meet the growing energy demands of industry and the popula-tion, Japan had to increase electricity generation quickly. However, it lacked natural resources and was increasingly dependent on foreign oil and gas. Since the 1960s, Japan's government has promoted an energy policy that favours the development of nuclear power as a secure and environmentally friendly (zero-emission) energy source.

The first nuclear power station in Japan started operating in 1963. By 2011, fifty-four commercial reactors were generating approximately one-third of the nation's electricity. These nuclear power stations were built in coastal areas, where they could be easily cooled by seawater and close to the energy demands of large population centres like Tokyo.

The area is no stranger to natural disasters. For example, in 1888, a volcano called Mount Bandai suddenly erupted so fiercely that it reshaped the local landscape, destroying three villages and killing some 500 people. Rocks and muddy debris from the explosions and landslides blocked the Hibara and Nagase rivers, creating the Bandai

Plateau with over 100 lakes, ponds and marshes of various sizes. Mr Tsurumaki, a monk staying nearby, gave an eyewitness account:

> Showers of large and small stones were falling all about us. To these horrors were added thundering sounds, and the tearing of mountains and forests presented a most unearthly sight, which I shall never forget while I live. It was pitch dark; the earth was still heaving beneath us; our mouths, noses, eyes, and ears were all stuffed with mud and ashes.

With whole villages destroyed and hundreds left homeless, it was a humanitarian crisis that shocked Japan. The fledgling mass media meant that photographs of the disaster quickly appeared in newspapers nationwide, and the response was overwhelming. Physicians from Tokyo rushed to the area, the Red Cross was mobilized and the new government immediately initiated a relief effort to help the victims. The Bandai eruption was a reminder of how vulnerable the Fukushima region was. Indeed, natural disasters such as volcanic eruptions, earthquakes and tsunamis are a common feature of the history of Japan

High up in the foothills of Aneyoshi, a small village on Japan's north-eastern coast, stands a ten-foot stone tablet carved with a warning to future generations: 'High dwellings are the peace and harmony of our descendants,' it says, 'Remember the calamity of the great tsunamis. Do not build any homes below this point.' Hundreds of these tsunami stones dot the coastline of Japan, giving testimony to the dangers of living near the ocean in an earthquake-prone area. Although some date back six centuries, many were built after a tsunami in 1896 killed over 22,000 people. Some call them the 'ancestor stones', and they warn people to drop everything if an earthquake strikes and to find high ground to survive. The word 'tsunami' is of Japanese origin, coming from the words for harbour, 'tsu', and wave, 'nami'. Japan sits on the western edge of the Pacific Ring of Fire, one of the most geologically active places on Earth, where tectonic plates are continually smashing into each other, creating volcanoes, islands and mountain ranges, and causing earthquakes. Mount Fuji, Japan's tallest mountain, is an active volcano that last erupted in 1707 and sits at a 'triple junction' where three tectonic plates interact. Some believe another eruption could happen at any time, and it is being closely monitored.

A stone tablet from the village of Aneyoshi which reads 'Home built high is children's relief. Remember the disastrous giant tsunami. Do not build homes below here' (T. Kishimoto via Wikimedia Commons)

Due to its unfortunate location, Japan experiences more earthquakes than almost anywhere else. Around 1,500 occur every year, but some are so small they are barely noticeable. The worst earthquake in Japanese history was the Great Kanto Earthquake, which hit the Kanto Plain surrounding Tokyo in 1923 and killed over 100,000 people. The earthquake that hit the city of Kobe in 1995 is still remembered worldwide for the scale of damage it caused, with more than 100,000 homes destroyed.

As Mark Twain said, 'History doesn't repeat itself, but it does rhyme.' The ancestor stones remind us that extreme earthquakes and the tsunamis that follow occur frequently in Japan. Since 1950, there have been 112 major earthquakes that also led to a subsequent tsunami. But their warnings seem to have faded over time. People are preoccupied with their busy lives, and many have put their faith in higher seawall defences installed by the government. Ignoring the advice of their ancestors, many places like Fukushima built homes and even nuclear power stations low down near the coast. 'It takes about three generations for people to forget. Those that experience the disaster themselves pass it to their children and their grandchildren, but then the memory fades,' says Fumihiko Imamura, a professor in disaster planning at Tohoku University.

The earthquake that struck on Friday, 11th March 2011, reached a magnitude of 9.1, making it the fourth most powerful earthquake in the world since records began. It knocked the Earth six and a half inches off its axis, moved Japan four metres closer to America and caused the coastline to subside by half a metre. The main quake was followed by over 5,000 aftershocks, the largest reaching magnitude 7.9.

It is hard to imagine the intensity and scale of the catastrophe, which came as a surprise to most people. Japan's main island, Honshu, lies at the intersection between three tectonic plates, but the epicentre of the 2011 quake was in an unusual spot, about 130 km east of Sendai, just off the northern coast – a subduction zone where the Pacific plate is being forced beneath the Eurasian plate. Strong earthquakes are possible here, but scientists hadn't thought there was enough energy to produce one larger than magnitude 7.5. A little over a year after the quake, the deep-sea-drilling vessel *Chikyu* bored deep into the subduction zone, taking temperature readings and collecting samples to help calculate the fault's friction and how much energy was released during the earthquake. They discovered that the Tohoku fault is lined with a thin layer of clay, making it much more 'slippery' than anyone thought possible. It slipped an unprecedented fifty metres, causing a rupture, which began deep underground, to reach the surface, where it triggered a sudden disturbance in the ocean and set off a tsunami.[1]

In the hours before the earthquake, there were reports of animals behaving strangely. People's pet dogs and cats were panick-
ing, biting their owners, barking loudly, hiding, being restless or
climbing high trees. In 2007, Japan's government spent $1 billion
on the world's most advanced earthquake early warning system,
with 1,000 seismographs scattered throughout the country. One
minute before the main shockwave hit, a warning alert buzzed on
the cell phone belonging to Professor Kensuke Watanabe. He told
Time magazine that he knew it was time for everyone in his class to
bolt under their desks. As the university building in Sendai began
to shake violently, Watanabe and his students were able to use the
sturdy desks as protection against falling objects. Shortly after,
they fled the building for open ground. Thanks to the warning, no
one was hurt. 'It was terrifying,' said Watanabe, 'but the mobile
warning really helped.'² The early warning system had never been
triggered before and automatically issued alerts via television and
cell phones shortly after the first, less harmful shock wave was
detected, providing time for many people to prepare for the more
powerful one that followed. Factories were able to shut down and
train services were automatically suspended. A string of detection
buoys in the Pacific Ocean picked up the tsunami, sending warnings to many nations, allowing enough time for people to switch
off their gas lines and get beneath a table or a door frame. It was
especially helpful to those in Tokyo, who were 230 miles from
the epicentre and had extra time to prepare. The towering tidal
wave had a maximum peak of 55 metres and took less than half
an hour to reach land. It flooded more than 470 square kilometres, destroying around 100,000 homes and forcing half a million
Japanese residents to be evacuated. Around 18,000 people were
killed, and it has since been estimated that it caused over $200
billion in damage, making it the costliest natural disaster ever.

While the earthquake caused violent shaking, making buildings
sway like rubber, it was the tsunami that exacted the greatest toll.
Coastal cities and towns were engulfed by a vast body of swirling water that swept away entire homes, left power lines gnarled,

threw up trucks and boats and ripped up trees. The majority of those killed in the disaster were victims of drowning. Entire towns were destroyed, and many of the survivors were left displaced and homeless. While survivors' descriptions of a tsunami differ, many recall hearing a sound of roaring, like a jet engine, getting closer and closer, seeing a horizontal white mist in the distance and smelling seaweed, mud and fresh fish. Tsunamis don't resemble the white-topped ocean breakers we are used to but are more like a surging tide – a wall of water that appears from nowhere and quickly swallows everything in its path.

Just before the tsunami reached the coastal city of Ishinomaki, forty-year-old plumber Kenichi Kurosawa scrambled three metres up a pine tree, wrapped his legs around a branch and held on for dear life. As the water rose to his knees, Kurosawa saw people in cars gripping their steering wheels as their vehicles were washed down the road. Others who had been hanging on to trees felled by the waves were swept away. For hours, Kurosawa endured sub-zero temperatures but survived. Afterwards, he told reporters, 'It's hard to imagine the power of a tsunami unless you've experienced it – it's a destructive force that just swallows everything up and obliterates everything in its path. I felt like the ocean was all around me. The water was so cold it chilled me to the bone.'

The journalist Richard Lloyd Parry spent six years reporting from the disaster zone and has written movingly about the tragedy and the impact it had on families and communities. In his book *Ghosts of the Tsunami*, he tells the haunting story of Okawa Primary School, the 'School Beneath the Wave', where seventy-four school children and ten teachers died when the tsunami inundated the town.[3] When the warnings came, Okawa's schoolteachers followed their emergency plan, which was written for fire, flood and epidemics, and evacuated to the nearby park outside the school. No one had ever seriously considered the possibility of a massive tsunami in this area. Despite the entreaties of parents and a few children to run for the hills and get on high ground, they stuck to the plan and stayed in the school's playground, the designated evacuation area.

Things were different in the tightly knit fishing community of Aneyoshi, where people sensibly built their homes above the ancestor stones: 'Everybody here knows about the markers. We studied them in school,' said Yuto Kimura, aged twelve. 'When the tsunami came, my mom got me from school and then the whole village climbed to higher ground.'[4]

The first wave to arrive at the Fukushima Daiichi Nuclear Power Plant was only four metres high and was deflected by the seawall, which was built to withstand waves of up to ten metres. Eight minutes later, a second wave hit, this time over fifteen metres high, breaching the wall. It destroyed the seawater pumps and flooded basements that housed the backup generators. Power was lost in five of the six reactors, and without a steady flow of cool water to the reactors' blazing-hot cores, meltdown inevitably followed. After a few hours, Prime Minister Naoto Kan declared a nuclear emergency, and the government issued evacuation orders for residents living near the plant. Over the next three days, inside the reactors, water evaporated and turned into steam, building pressure that created leaks and allowed radioactivity to escape. Workers 'vented' the containments to try to reduce the internal pressure, releasing yet more radiation. As the reactors overheated and fuel melted, highly flammable hydrogen was generated, causing a series of massive explosions. Immediately afterwards, radiation levels near the plant were measured at 400 millisieverts per hour. By comparison, the average person is exposed to about 2.4 millisieverts of radiation per year, meaning that the radiation immediately surrounding Fukushima was 1.46 million times stronger than it would be in an average environment. Emergency workers continued to fight to cool the plants, pumping in seawater, and Japan's Self Defence Force used helicopters to drop seawater directly on to the core of one of the reactors. Things began to stabilize and electrical power was eventually restored eleven days after the crisis began, but not before significant amounts of radiation had been released into the atmosphere and the Pacific Ocean.

Concerns over radiation exposure led the government to evacuate everyone within a 20 km zone around the plant, and over 160,000 people were relocated, many of whom have never returned. The true human and environmental impact of the meltdown has taken experts years to assess, but at the time the biggest concern was increased radiation levels in food and water. The radioactive fallout included many volatile radioactive isotopes, such as iodine-131, caesium-134, caesium-137, and more xenon-133 than was released at Chernobyl. These radioactive chemicals are carried by air and water, seep into the soil and contaminate crops or get transferred to animals, and so enter the food chain. This can build up over time and have widespread effects. Tuna as far away as California were found to contain low levels of radioactive caesium from Fukushima. If you visited the nearby town of Okuma ten years later, you would see rain-soaked fields piled high with thousands of black sacks containing radioactive soil. Okuma was still too radioactive for residents to return and instead was turned into a processing plant for the topsoil, grass, tree branches and other material that was removed from areas near homes, schools and public buildings around Fukushima in the years following the disaster. The biggest nuclear clean-up the world has ever seen has generated millions of cubic metres of soil packed in bags and neatly stored in these temporary facilities. The government has promised to find somewhere outside Fukushima prefecture to finally store the material, but this poses a conundrum: no one wants it in their backyard.

Although around 122,000 of the people evacuated have now returned to their homes, anxieties over the cumulative effect of radiation over a long period of time still linger. So far, very few people have had health problems due to radiation, and the Japanese government and international experts have gone to considerable lengths to reassure the public that the region is now safe. But the worst health effects were not due to radiation. It turned out that immediate radiation fallout from the meltdown was trivial, and residents living near the plant were not exposed to anything approaching harmful levels. But the evacuation itself, designed to protect people, caused so much trauma that one study found 59.4 per cent of those who were displaced were found to have post-traumatic stress disorder (PTSD).[5] This was connected to chronic physical diseases, worries about livelihood, lost jobs, lost social ties and concerns about compensation. There was an increase in diabetes due to problems accessing medical care, as well as obesity because people were afraid to go outside and exercise. Mental health problems were widespread, often leading to depression and suicide. Mothers who lived near the nuclear plant were found to have high levels of depression – suffering from anxiety over the potential effect of the radiation on their children. They worried about the safety of their children's food, about going outside, and their financial situation. Overall, stress, interrupted medical care and suicide caused an estimated 2,202 deaths among the evacuated population, particularly among elderly people in temporary housing.

Risk can never be eliminated, and natural disasters strike with tragic frequency. But activities like nuclear power generation, which has the potential for cataclysmic impact if things go wrong, need to put understanding risk at the heart of their design and planning. Listening to the past and imagining all possible future scenarios is critical. The operator of the Fukushima plant, the Tokyo Electric Power Company (TEPCO), has been heavily criticized for their corporate culture and for not being up to date with international best practices. Had they followed the latest international standards, the accident could have been prevented. Evidence that massive tsunamis hit the area about once every thousand years should have been taken into account. At the time, European counterparts were generally much better protected against flooding, and if state-of-the-art safety approaches had been adopted, the worst effects could have been avoided.

'The earthquake and tsunami of 11th March 2011 were natural disasters of a magnitude that shocked the entire world,' said Kiyoshi Kurokawa, the chairman of the Fukushima Nuclear Accident Independent Investigation Commission. 'Although triggered by these cataclysmic events, the subsequent accident at the Fukushima Daiichi Nuclear Power Plant cannot be regarded as a natural disaster. It was a profoundly manmade disaster – that could and should have been foreseen and prevented.'

Just 60 km along the coast is the Onagawa nuclear power station. It was much closer to the epicentre and experienced the worst impact of any of the nuclear facilities in the area, in fact experiencing stronger shaking than any nuclear plant has ever had from an earthquake. The tsunami was also bigger at Onagawa, yet despite this, the plant shut down safely and was remarkably undamaged by the disaster. Both Fukushima and Onagawa were built low down on the coast, so how can we explain their differing fates? The answer is simple: Onagawa's builders had committed themselves to the concept of *failing safely* if anything ever went wrong. They studied the history of the area and all the possible risks, including the likelihood of a tsunami, and made that a central part of the entire design. They carried out surveys and simulations to predict likely tsunami levels and built their reactors at a higher elevation, 14.7 metres above sea level, five times the average tsunami height. Importantly, they also regularly updated this research, kept an eye on other earthquakes and tsunamis around the world and continually improved their safety measures.

Fukushima, on the other hand, significantly underestimated the tsunami risk and built much lower with fewer precautions in the design. Costas Synolakis, director of the Tsunami Research Center at the University of Southern California, described their approach as a 'cascade of stupid errors that led to the disaster'. Led by its *failing safely* approach, Onagawa also invested heavily in its emergency response capability and safety culture, carrying out extensive scenario-planning exercises and ensuring staff were well-trained in emergency procedures. Unusually for a Japanese organization, they also devolved responsibility to the local supervisors and chief engineers, empowering them to make decisions in the moment should a crisis occur. Most people have never heard of the Onagawa plant, yet it is one of the best examples of a risk mindset and safety culture you will find.

Speaking three weeks after the accident, the Japanese Prime Minister Naoto Kan showed that the failing safely philosophy had been understood at the highest levels:

> We must prioritize the health and safety of the people of Japan… we must implement risk-management initiatives to such an extent that some in the public feel we are being too cautious… and we must conceive of every possible scenario and prepare response systems that can deal with each scenario should it occur.

Sayonara Nuclear Power

Visited by millions of people every year, the Meiji Jingu Shrine in Tokyo is hidden in a densely forested area a stone's throw from the quirky streets of Harajuku – a trendy hub for teenage fashion and cosplay culture in Japan's busy capital city. Built in 1915, it commemorates Emperor Meiji, whose forty-five-year reign saw Japan transformed from a feudal state to an industrialized world power. It is one of Japan's most popular Shinto shrines, and visitors traditionally take part in offerings at the main hall, buy charms or write out their wishes on an *ema* – a wooden plate where visitors leave their wish cards in the hope that they will come true.

The protesters who gathered here on 19th September 2011 wished for one thing: the immediate closure of all of Japan's nuclear power stations and a new energy policy focused on renewables. Chanting 'Sayonara nuclear power' and waving 'No nukes' banners, over 60,000 people marched in a remarkable demonstration of civil feeling in a country where public protests are almost unheard of. Nobel laureates, musicians and actors joined residents of Fukushima prefecture at the rally, shouting: 'No more nuclear power plants! No more Fukushimas!' Keiko Ochiai, a well-known writer, joined Nobel laureate Kenzaburo Oe at the head of the march. 'We believe that nuke is the reason for the unhappiness of the human beings,' she said, 'so I can't agree with nuclear or nuclear plants.' One of the protesters, Kazuhiro Hashimoto, a medical-service employee from Fukushima, told journalists: 'If we fail to abolish nuclear power plants now, we will never achieve a nuclear-free world. It's too late for us to raise our voices after another nuclear accident occurs. We hope the Fukushima accident will be the last one.'

It was the first in a growing anti-nuclear movement in Japan and bigger protests followed. It signalled an almost complete loss of faith in the government, whom the public blamed for the disaster, and in an energy policy that many believed was one of the cleanest and most progressive in the world. Immediately after the Fukushima meltdown, all but two of Japan's fifty-four nuclear power stations were shut down or suspended pending safety reviews. The energy shortage this created caused consternation for the Prime Minister and his government, who were worried about how they were going to maintain electricity supplies. They were forced to resort to a drastic conservation programme that included turning off air-conditioning during the summer and office lights during the day. Since then, just ten plants have been allowed to restart operations, with another twelve due to come back online by 2025, and a further eighteen by 2030. In 2017, the government said that, if the country is to meet its Paris Climate Accord targets, nuclear energy would need to make up roughly twenty per cent of its energy mix.

The disaster did little to alter the path of most other countries. Those that were committed to boosting nuclear power stuck to their plans, and those that favoured phasing it out continued to do so. Only Taiwan, Belgium and Germany were prompted to speed up their plans to shut down nuclear reactors. The biggest reaction came from Italy, where a public referendum on nuclear power two months after Fukushima returned a resounding 'No!', with 94 per cent voting against the Italian Prime Minister Silvio Berlusconi's plans to resume nuclear power generation.

Fukushima caused a rupture in Japanese society and shook the foundations of a culture that traditionally affords great respect to those in authority or credentialled experts. Surveys conducted after the accident showed that public trust had been totally shattered.

They blamed the companies who operated the nuclear power plants as well as regulators who had failed to keep them safe. Confusion, contradictory advice and poor-quality information in the immediate aftermath, such as issuing evacuation orders and communications on food safety, caused trust in both government and nuclear experts to plummet. Moreover, the public felt duped by those in authority who had perpetuated a 'nuclear safety myth' – that nuclear power was completely safe and an accident would never happen in Japan. Fear of unknown catastrophes and radiation anxiety were heightened for years afterwards, making the public discourse even more complicated. While the public reaction to Fukushima is easy to understand, it highlights an important issue in how we evaluate risk. To make good decisions, we need to understand as much as possible about all the risks we face, especially when some are interconnected or have implications we might not easily foresee. With risks, there are always trade-offs. A study published in 2019 showed that the closure of nuclear power stations following Fukushima led to more deaths than the accident itself. Nuclear power was replaced by fossil fuels, causing a huge increase in electricity prices, which led to a reduction in energy consumption. As a result, more people died during the freezing winter, and there was a spike in respiratory disease due to the increase in the burning of fossil fuels.[6]

Japan is now slowly returning to nuclear power. The global energy crisis, rising heating bills over the winter and the prospect of blackouts during the summer are causing people to re-evaluate the benefits of cheaper, more secure energy. Supporters of nuclear power have overtaken opponents for the first time since Fukushima. A survey by Asahi Shimbun newspaper in March 2023 found that 51 per cent of people in Japan now support the restart of nuclear power stations, compared to 42 per cent against. The current Prime Minister, Fumio Kishida, is positioning the nuclear revival as part of a 'green transformation' and has committed to 'constructing next-generation nuclear reactors equipped with new safety mechanisms', as well as 'making maximum use of existing nuclear plants'. While support is growing, Fukushima is still strong in the public memory, and rebuilding trust will not be easy – the government, regulators and energy companies will need to show they have changed.

What have we learned from Fukushima? The evacuation of people from the suspected radiation zone was so badly handled that it exacerbated the crisis and led to pain and suffering that could have been prevented. It has taught us that we must be cautious about large-scale evacuations and pay more attention to the consequences they have on health, especially with the sick or elderly. If an evacuation is needed, proper medical support and supervision must be provided. During the earliest stages of the crisis, communication between the government and people living near the reactor was chaotic and misguided, adding significantly to the panic caused by the news. Understandably, it undermined public confidence in the government and experts associated with nuclear power for years to come. Much has been written since about how best to communicate with the public during a meltdown in a nuclear power station. But perhaps the most obvious lesson to be learned is to avoid building nuclear power stations in seismically active areas, or at least to build them high up, away from coastlines. 'Do not build below this point,' said the ancestor stones. Naoto Kan, who resigned as the Japanese Prime Minister just a few months after the accident, outlined his vision for the future:

We must then begin preparations toward reconstruction. In fact, we will go beyond mere reconstruction, creating an even better Tohoku and even better Japan... We are moving forward with the creation of a reconstruction plan that has this big dream at its core... in some areas we will level parts of mountains in order to create plateaus for people to live on. Those residing in the area will then commute to the shoreline if they work in ports or the fisheries industry...

The wisdom of the ancestor stones would be at the heart of Japan's new philosophy. History gives valuable insight into the kind of events one can anticipate in the future, which is helpful in evaluating risk.

As well as listening to the past, we also need to harness the power of imagination to work out what the future might have in store. As Onagawa's success showed, good crisis planning for high-risk sites like a nuclear power station should consider all possible future risks and plan for worst-case scenarios. If this approach had been adopted in the design and management of Fukushima, its builders would have been prepared for the effect of the tsunami and ensured that backup generators were kept in watertight rooms high up, away from possible flooding. The impact of climate change, terrorism, our ageing population and interconnected economic crises are making the world more uncertain than ever before, and it is harder to predict the future. Foresight isn't a perfect discipline, but there are internationally recognized techniques – scanning the horizon for emerging changes, studying local and global trends and testing multiple scenarios or 'war games'. Better foresight is needed to make our critical infrastructure safer and our communities more resilient. To do this, we need to listen to a large and diverse group of experts' advice and involve the public in the process. Public participation gives unique perspectives often overlooked by experts with deep subject matter knowledge and helps improve public communication and, ultimately, trust and acceptance.

After the tragedy at Okawa Primary School, authorities throughout Japan pledged that never again would a school be poorly prepared for a tsunami. Evacuation drills have become a regular part of school life, emergency plans have been updated, new equipment has been provided, and everyone knows what to do and where to go if disaster strikes. Japan now takes a leading role in tsunami preparedness in the Asia-Pacific region, working with the United Nations Development Programme to raise awareness and provide training in schools that could be at risk. Since 2017, over 160,000 students, teachers and administrators from 330 schools in twenty-three countries have been trained in tsunami preparedness. World Tsunami Awareness Day takes place on 5th

November, a date chosen to recall the story of a Japanese village leader who, in 1854, set fire to rice sheaves to alert the population to an imminent tsunami and save lives. Approved by the UN General Assembly in 2015, the awareness day first took place in 2016, when 360 school children from thirty countries around the world gathered in the Japanese coastal city of Kuroshiro for a youth summit on tsunami risk. Kuroshiro is forecast to be in the path of a thirty-four-metre high wave if an earthquake occurs in the Nankai trough off Japan, so tsunami preparation is particularly poignant for the local community. During the two-day event, students held discussions, attended briefings and undertook fieldwork. They looked at what they could do for their communities to be better prepared and more resilient to disasters and tried to learn from each other by listening to different experiences. The summit was chaired by a fifteen-year-old Kuroshio high-school student, Ren Imai, who said: 'After having listened to students who had their own tsunami experience, I want to learn more about disasters caused by tsunami so that I can enhance my understanding and awareness towards the future.' They concluded the summit by issuing the Kuroshiro Declaration, in which they vowed to spread awareness in their communities about tsunami threats and how to tackle them, and were formally designated as Youth Ambassadors for Tsunami Awareness. After Fukushima, there is hope that the next generation will not ignore the lessons of history.

Chapter 4

A Destroyer and a Teacher

Increased global travel, urbanization and the rapid exploitation of the natural world over the past century have all increased the likelihood of pandemics, and these trends are set to accelerate. Covid-19 was the most recent reminder that sudden, large-scale infectious disease outbreaks are one of the biggest risks shared by everyone across the planet. When new pandemics emerge, the combination of terrifying symptoms, fear of the unknown and high death rates seem to grip the imagination and cause widespread panic amongst the public. The latest statistics from the WHO indicate that from 2019 to 2023, there were 772 million confirmed cases of Covid-19 and approximately seven million deaths worldwide.[1] While Covid-19 isn't the deadliest pandemic of all time, it is the most widespread and has affected almost every nation in the world (except for a few small island states in the Pacific and Atlantic), making it the first truly global pandemic. Human history is full of examples of the devastating impact of fast-spreading disease – from the Plague of Justinian that heralded the end of the Roman Empire and the bubonic plague that ushered in the 'Dark Ages' to more recent epidemics such as cholera, Spanish flu and HIV/AIDS. Why weren't we better prepared for Covid-19? New research shows that pandemics are happening more often and now travel so fast that they are almost impossible to contain. So, what can we learn from past pandemics and how prepared are we for the next one?

The earliest recorded pandemic, the Plague of Athens, ravaged Greece, North Africa and the Middle East between 430 and 426 BC, resulting in just under 100,000 deaths in Athens alone, around one-quarter of the city's population. The cause is unknown, but many candidates have been suggested, including typhus, smallpox, measles or possibly even Ebola. Due to the Peloponnesian War, Athens was severely overcrowded at the time, and people had little practical understanding of the importance of hygiene in fending off infection, which meant the disease spread like wildfire. It was well documented by the Athenian historian and general Thucydides in the *History of the Peloponnesian War*. He describes the symptoms, which included a terrible fever:

> People in good health were all of a sudden attacked by violent heats in the head, and redness and inflammation in the eyes, the inward parts such as the throat or tongue becoming bloody and emitting an unnatural and fetid breath.

Anyone who tried to help quickly became ill themselves, and so people abandoned the sick, even family members, to stay safe. In Athens, no one had ever experienced this disease before. Still, people knew that quarantining those who fell ill, either voluntarily or by force, effectively contained its spread. Civil leaders ordered anyone with symptoms to stay home, and some willingly barred themselves inside their houses to avoid infected people. But any authority the democratic government had quickly evaporated after the death of Pericles, the ruler and first citizen of Athens. The plague had a devastating impact on Athenian social order, which completely broke down. According to Thucydides, 'the catastrophe was so overwhelming that men, not knowing what would happen next to them, became indifferent to every rule of religion or law.' Panic and hysteria took over, and people lost all fear of the law because they felt they were living under a death sentence. There are some similarities with Covid-19 in how the constant state of shock and panic during its early stages was demoralising, and 'lockdown fatigue' brought on apathy and a lack of commitment to safety measures.

Derived from ancient Greek, meaning something belonging to all people, the word 'pandemic' was first used to refer to the outbreak of a disease in 1666, the same year as the Great Plague of London, and referred to 'a disease always reigning in a Country'.[2] Until the nineteenth century, pandemics were largely unexplained and uncontrollable. There was very little knowledge of disease mechanisms and poor public healthcare, so disinformation, 'fake news', and the use of bizarre treatments with no scientific basis were rife. This changed with the emergence of the science of modern statistics, which could be applied to health problems with life-saving results. One of the first to do this was Florence Nightingale, who studied mathematics before training as a nurse and had a lifelong passion for statistics. As a young child, she was a prodigious collector of seashells and enjoyed the process of cataloguing them, creating lists and tables, and bringing order to nature to understand it better. She pioneered the consistent recording of sickness and death records in military hospitals during the Crimean War, gaining an insight into the numbers and causes of death. This led to the remarkable discovery that soldiers were seven times more likely to die in hospital (due to unsanitary conditions) than in combat. Nightingale used her knowledge of statistics to develop an entirely new way of presenting this data visually to convince the British Parliament to act. This statistical approach went beyond just counting the numbers and causes of death. It provided the basis for a new type of risk analysis that was able to unravel the underlying chain of events that lead to outbreaks of disease.

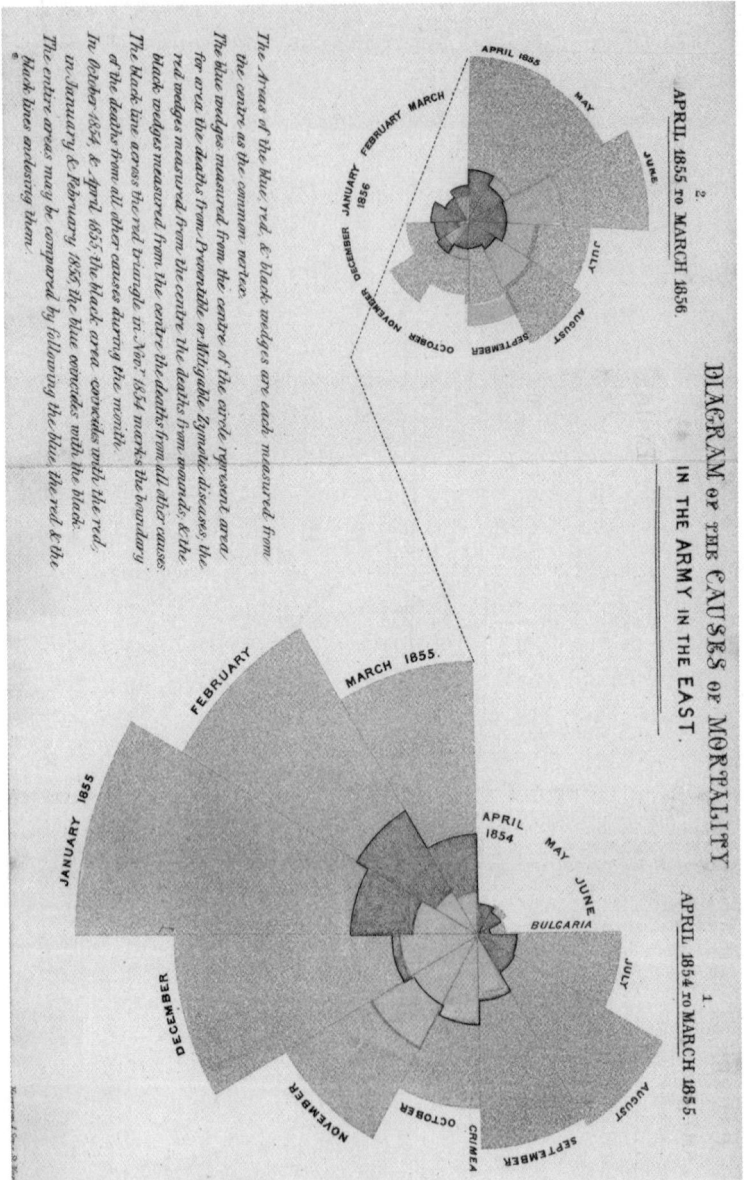

Florence Nightingale's *Diagram of the Causes of Mortality in the Army in the East*. Published in *Notes on Matters Affecting the Health, Efficiency, and Hospital Administration of the British Army* and sent to Queen Victoria in 1858

In the late eighteenth and early nineteenth centuries, Britain, which was already the world's leading commercial nation, was reshaped by the Industrial Revolution. Ingenious new machines meant that textiles could be produced in vast quantities, the steam engine transformed the productive capacity of factories and the telegraph made the world a smaller place. Millions left the country-side and sought out newly created jobs in Britain's rapidly growing industrial cities: Leeds, Birmingham, Liverpool and, most of all, London. With a population of just one million in 1800, London was the heart of Britain's innovation engine. Offering the most fantastic range of jobs, people flooded to the capital, which swelled to become the world's largest city (reaching a size of 6.7 million by 1900). While there were jobs aplenty, there was little housing, and most had to live in overcrowded slums. The city soon became overwhelmed by this population explosion, and human waste piled up in the streets and cesspools overflowed into gutters and waterways. There was no plumbing, and few had access to run-ning water. All toilet waste, animal carcasses or butcher's remains and general garbage were carried into the river Thames, which became so polluted the stench was unbearable. These squalid con-ditions were the perfect breeding ground for disease, and typhoid and scarlet fever were commonplace. Charles Dickens described the Thames in *Little Dorrit*: 'Through the heart of the town a deadly sewer ebbed and flowed, in the place of a fine fresh river.' In 1831, a terrifying new disease emerged in the slums of London with strange symptoms never before encountered by European doctors. Known as Asiatic cholera, it was a 'foreign invader' that caused uncontrollable diarrhoea and vomiting, resulting in severe dehydration from fluid loss. There was no cure, and no one knew how it spread.

Cholera is an ancient pandemic that persists to this day, affect-ing people in poor communities who do not have access to safe water or good sanitation, mainly in Africa and Asia, where it takes advantage of breakdowns in infrastructure caused by natural dis-asters or conflict. According to the WHO, the world is currently experiencing an acute upsurge in cholera cases, with over thirty countries reporting outbreaks in 2022. It has existed in India for at least a millennium, and the people of Bengal even have a goddess, Oladevi, whom they worship in the belief she will protect them from cholera and other stomach illnesses. In the nineteenth century, if there was an outbreak, villagers would gather, and the village head would lead a ceremony of devotion. At the time, cholera was one of the leading causes of death amongst British seamen in Calcutta, whose harbour was described in the *Indian Medical Gazette* as 'the Maelstrom of death'. It spread from Bengal in 1817, carried in the bilge water of East India Company ships and by seafarers themselves, to become the first global cholera pandemic. Initially, it was concentrated along the trade routes that connected the Indian subcontinent to Asia, the Middle East, and Europe and spread by soldiers being transported to postings throughout the British Empire. The second pandemic began ten years later in 1826, and this time spread much more widely, with severe outbreaks affecting the major cities of Russia, Great Britain and the Americas.

It first appeared in Britain in 1831, with further outbreaks in 1837 and 1838. Although poor sanitation and disease were rampant in Victorian London, it was the final straw for the liberal Whig government, which asked the lawyer Sir Edwin Chadwick to carry out a public inquiry into the problem. Chadwick, a disciple of utilitarian philosopher Jeremy Bentham (who believed in 'the greatest good for the greatest number'), was a champion of the poor and campaigned for new laws to help alleviate poverty in Britain. His report, *The Sanitary Conditions of the Labouring Population*, was published in July 1842 and was one of the most critical developments in nineteenth-century social reform. It investigated urban centres across Britain and took a unique quantitative approach, using hard statistical analysis alongside environmental descriptions. It revealed a direct link between poor living conditions, disease and life expectancy, fuelling calls for an urgent national response. As Chadwick put it:

Britain's dire urban environments and the ill health it helped promulgate was akin to a war that was killing more every year than any military conflict in which Britain had ever been involved.

It was a damning indictment of society that caused widespread public debate and led to the 1848 Public Health Act that set out for the first time the responsibility of the state to protect the health of its citizens. It advocated for better sanitation, in particular measures such as cleaning, better drainage, rapid removal of human waste and good ventilation. Chadwick had identified a connection between poor sanitation and cholera but not the cause, hampered by the medical thinking of the day that believed infectious disease was transmitted by noxious air called miasma. In miasma theory, disease was caused by poisonous vapour containing particles of decaying matter that caused a foul smell. While better housing, cleaner streets, and improved public hygiene had a positive effect, and levels of disease fell, they didn't eliminate the underlying cause of cholera, so it returned in 1848 and 1853.

A new kind of statistical analysis was needed to get to the bottom of the problem. British physician and mathematician Sir John Snow, a vegan teetotaller from York, believed that cholera was not carried by bad air but by contaminated water and went about trying to prove it. One of nine children, Snow grew up in one of the poorest neighbourhoods in the city of York, frequently flooded by the river Ouse and contaminated by sewage, and so had first-hand experience of the impact of inadequate sanitation on a community. He showed an early aptitude for mathematics and was lucky enough to obtain a medical apprenticeship to become a surgeon-apothecary, working in Newcastle before moving to Westminster Hospital in London, where he graduated in 1844. As a young apprentice, he experienced a cholera epidemic in a coal-mining village, and the trauma of his early life led him to vow to resist drink, gambling and marriage. Snow was ahead of the times in many ways, becoming a vegetarian at the age of seventeen, then later a vegan, before his health deteriorated and he suffered from kidney disease. He thought this was due to his diet, and so, in the interests of his health, he took up eating meat and drinking wine. He created new techniques in surgical anaesthesia, becoming the personal anaesthetist to Queen Victoria, but it is for his public health work that he is most remembered, founding the modern science of epidemiology and the underpinning principles of risk analysis.

From August to September 1854, there was another cholera outbreak, this time concentrated in the Soho area of London near Broad Street, an area characterized by overcrowded housing and poor hygiene. The disease had already driven Soho's wealthier residents to move to the cleaner, more open spaces of Mayfair and Belgravia. Snow investigated and found that 600 deaths had occurred over a ten-day period. Plotting the numbers of deaths on a map of the area, he noticed that most cases were clustered near the Broad Street pump, and all of them had used this for drinking water. Nearby brewery workers and poorhouse residents who took their water from local wells instead had escaped the epidemic, providing further evidence that the source was the Broad Street pump. Indeed, the 535 workers at the Poland Street Brewery seemed remarkably immune, with only five contracting cholera. This was because the brewery allowed them to drink as much of the malt liquor they made as they wanted. They existed almost exclusively on free beer, never touching the water from the pump outside – proof that drinking beer is good for you. Snow met with the Board of Guardians of St James's parish and convinced them to remove the handle of the pump to disable it. Within a few days, new cholera cases in the area had vanished. It was later discovered that the water for the pump was contaminated by sewage from a leaking cesspit nearby.

Monster Soup by William Heath, 1828
(Wellcome Library, London)

It was well known that the Thames was severely polluted, and drinking water from the river was popularly called 'Monster Soup'. Indeed, in 1834, the Anglican cleric and wit Sydney Smith said, 'He who drinks a tumbler of London water has literally in his stomach more animated beings than there are men, women and children on the face of the globe.' Snow also carried out a study of water being supplied to residents of south London, showing that the company who pumped water from sewage-polluted sections of the Thames caused a much higher incidence of cholera. Yet despite all this, Snow's research wasn't enough to convince the General Board of Health that cholera was spread by contaminated water. It wasn't until London's final outbreak in 1866 that the link was conclusively proved and finally accepted by the authorities. By developing new ways to look for patterns in the data, including one of the first applications of geospatial visualization, Snow's pioneering research successfully traced the mode of transmission to the water supply (even though the cholera bacterium would not be discovered until 1884) and transformed the world's approach to public health. In doing so, he founded the field of epidemiology – the systematic study of the distribution, patterns and causes of disease in a population. Unlike his predecessors, Snow went further than simply tallying mortality statistics. He used his own theories about the transmission mechanism, together with careful observation and innovative analysis, to identify the underlying cause. Moreover, he used his insights to convince the parish authorities to take action that eliminated the danger. This is the basic principle that underpins the modern field of risk analysis, where experts study health problems, try to identify possible causes and use this information to reduce the danger and convince policymakers to make permanent changes. A replica of the pump (still with no handle) was installed in 2018 as a monument to Snow on Broadwick Street, outside the pub which now bears his name. The disinfection of drinking water by filtration and chlorine has provided clean water to the world. It is one of the most significant public health achievements of all time, eradicating cholera and other water-borne diseases like typhoid fever and dysentery in almost every country.

The Broad Street Map from *On the Mode of Communication of Cholera* by John Snow. Cholera cases are highlighted in black, showing the clusters (indicated by stacked rectangles) in the London epidemic of 1854.
(C.F. Cheffins, London, 1854 via Wikimedia Commons)

In Flew Enza

Largely thanks to advances in our understanding of viruses and how to treat them, the Covid-19 death toll was lower than many historic pandemics. The deadliest pandemic in recorded history was Spanish flu (or H1N1 Influenza A) in 1918–1920, which came from nowhere and killed more people than the First World War. By 1918, the war was entering its fifth and final year, and the Allied armies were winning, thanks to a new approach under the French General Ferdinand Foch, whose 'blow-by-blow' strategy was working. This method of coordinated attacks at different parts of the front was 'war-winning military art'[3] and, together with an injection of American troops to replace French losses, the tide had finally turned in the Allies' favour when an unexpected enemy struck – a seemingly common flu that soon ravaged armies on both sides. It was the very need to recruit huge numbers of new troops to win the war that created the spark that allowed the new pandemic to take hold. Shortly after the USA declared war on Germany on 6th April 1917, Congress approved the colossal sum of $3 billion to build a 'million-man army'. President Woodrow Wilson appointed General John 'Black Jack' Pershing to build and lead the American Expeditionary Force, which was only about 27,000 strong at the time. Black Jack immediately set about a massive recruitment drive, setting up over thirty new training camps to handle the hundreds of thousands of Americans drafted into the army. It was in these hastily established military camps that the first outbreak began in March 1918 – overcrowding and unsanitary conditions created a fertile breeding ground for the new virus. It had a particularly short incubation period, with people becoming symptomatic in just a few days (compared to five to six days for Covid-19), and so it spread rapidly. By the summer of 1918, approximately two million US troops had arrived in France, incubating the virus during the long journey across the Atlantic, where it exploded through the armies of both sides and quickly took hold in Europe.

It became known as 'Spanish' flu because the first newspaper reports came from Spain, where bulletins were not censored, unlike in Germany and Allied countries, which went to great lengths to conceal the impact of the pandemic on the strength of their armies. It had a devastating effect on Spain, which was still an agrarian economy. The majority of the population lived in poverty, with little education and in generally bad health. Mortality rates in Spain were the highest in the developed world, and the poor were hit hardest because they couldn't afford medical care or to take time off from work to self-isolate. Spanish authorities recommended washing hands often, avoiding crowded places, cancelling festivals, and closing theatres and schools. However, unlike in other countries, wearing masks and workplace closures were not implemented, making self-isolation a personal choice that only the rich or those with personal savings could afford. More than a quarter of a million people died, approximately 1.25 per cent of the Spanish population. Shockingly, the mortality rate amongst those with higher incomes (liberal professionals and property owners) was 29 per cent, compared to a staggering 69 per cent for those with low incomes. Given the state of the Spanish economy, survival was largely dependent on one's own financial means, leaving a long legacy of inequality in Spanish society.

In 1918 there were no vaccines or treatments for flu, and medics had very little knowledge of how to handle an outbreak. It baffled scientists from the outset because it mainly attacked young and apparently healthy adults between the ages of twenty and forty, whereas previous flu epidemics had only seriously affected the elderly or the sick. Around 500 million people, approximately one-third of the global population at the time, were infected. The lack of good medical records means no one knows the exact death toll, but estimates range from 17 to 100 million people worldwide. Not only was it incredibly infectious, but it was also deadly: 2.5 per cent of those infected died, compared to less than 0.1 per cent for a normal flu epidemic. A rare genetic mutation of the flu virus that appeared out of nowhere, it was totally new to most people in the world, and so there was no herd immunity to protect the population. The science of immunology was in its infancy, and while scientists and pharmaceutical companies raced to try to find a vaccine, they didn't yet have the ability to respond fast enough. The onset of the illness was incredibly swift – people perfectly fine and healthy at breakfast could be dead by dinner time. The first symptoms were generally headache, fever and tiredness, and some people would quickly develop pneumonia and start turning blue, struggling for air until they suffocated to death. Doctors and nurses, mostly from the Red Cross, were overwhelmed and worked day and night to help sufferers, but with no effective treatment or antibiotics there was little they could do except try to alleviate the worst of the symptoms.

As the war came to an end, soldiers returning home brought the virus with them, and the second and third waves of the pandemic hit in the winter of 1918–19, bringing America and the UK to a standstill. There was an almost comprehensive closure of schools, theatres and public gatherings, and most families lost loved ones. Cities were hit hardest – New York buried 33,000 victims; undertakers struggled to cope and ran out of coffins, forcing family members to bury their own dead. In the streets (if they were allowed out), children chanted a cautionary skipping rope song:

> I had a little bird
> Its name was Enza,
> I opened the window
> And in flew Enza.

This rhyme was popular during the 1918 pandemic, but originated from an earlier song most likely written during the forgotten flu epidemic of 1889–90:

> There was a little girl, and she had a little bird,
> And she called it by the pretty name of Enza;
> But one day it flew away, but it didn't go to stay,
> For when she raised the window, in-flu-Enza.

Like radio advertising earworms, it was a hugely effective way of communicating a common risk, but was based on the mistaken belief that infections were carried by 'poisonous' air and that people needed to keep their windows closed to keep it out. Health campaigns at the time advised the total opposite to reduce the risk of spreading infection in confined spaces. In Chicago, police officers were ordered to arrest anyone sneezing or coughing in the street, as 'Coughs and Sneezes Cause Diseases'. Massive public health campaigns promoted the best available information and advice in an attempt to keep people safe. Cartoons, streetcar signs, posters and newspaper adverts warned people not to spit in public – 'Spit

Spreads Death' – to keep bedroom windows open, and to wear a mask – 'Wear a Mask or Go to Jail'. In many places, masks were made mandatory, and some considered it a patriotic duty to wear one. Newspapers printed instructions on how to make your own mask at home and anyone who didn't wear one was publicly shamed and branded a 'Mask Slacker' – a reference to the men who didn't participate in the war effort. Officials even released a musical jingle to advocate for mask-wearing:

Obey the laws, and wear the gauze.
Protect your jaws from septic pause.

It affected everyone; Woodrow Wilson was suffering from it in early 1919 while negotiating the Treaty of Versailles. Other survivors included Walt Disney, British Prime Minister David Lloyd George, Mahatma Gandhi, Greta Garbo and the painter Edvard Munch. Munch was moved to paint two self-portraits, one during the flu and one afterwards, showing his experience with the illness and its legacy. Munch depicts himself sitting in a wicker chair in front of an unmade bed, a muted colour scheme conveying a sense of melancholy and existential drama reflecting the trauma and despair universally felt at the height of the pandemic. Physician George Price wrote that Spanish flu was both 'a destroyer and a teacher'; indeed, it taught us a great deal about both communicable diseases and how to influence public behaviour in a new era of mass media.[4] In 1918, people quickly realized the pandemic was mostly spread in crowded places or where people gathered in close proximity. Recent research shows that early and sustained social distancing measures, such as bans on gatherings and the closing of public buildings, prevented a much higher death toll. The public response to these measures and to the advice of authorities was remarkably calm and obedient. In the aftermath of the Great War, during which the government imposed harsh measures such as rationing and the draft, the public had become inured to the loss of some personal freedom.

Perhaps as a result, governments and health officials found it much easier to enforce strict measures and advice than they did during either the Plague of Athens or Covid-19.

TREASURY DEPARTMENT
UNITED STATES PUBLIC HEALTH SERVICE

INFLUENZA

Spread by Droplets sprayed from Nose and Throat

Cover each COUGH and SNEEZE with handkerchief.

Spread by contact.

AVOID CROWDS.

If possible, WALK TO WORK.

Do not spit on floor or sidewalk.

Do not use common drinking cups and common towels.

Avoid excessive fatigue.

If taken ill, go to bed and send for a doctor.

The above applies also to colds, bronchitis, pneumonia, and tuberculosis.

US Treasury Department public health poster with instructions for preventing the spread of influenza. Distributed during the influenza pandemic of 1918 (United States Public Health Service, Treasury Department, 1918)

Edvard Munch, 1919, *Self-Portrait with the Spanish Flu*
(National Museum of Art, Architecture and Design, Oslo)

The Covid War

Spanish flu came to an end in the summer of 1919, as those who survived developed immunity. In 2008, researchers discovered a mutation in a group of three genes that made it so deadly – it enabled the virus to attack a victim's bronchial tubes and lungs and make them susceptible to bacterial pneumonia. There were many lessons to be learned from the deadliest pandemic ever, but it was hardly a priority as the world struggled to recover from the social and economic aftermath of the First World War. One hundred years before Covid-19 struck, we had a good understanding of how pandemics spread and how to protect the population. We also had the analytical tools to identify pandemic risks and discover the underlying causes. Previous pandemics have taught us the importance of self-isolation or quarantine, face masks, good hygiene and mass-media campaigns to spread accurate and timely information. Despite all this, Covid-19 came as a shock to both the world's citizens and governments – the public didn't know what to do, and poorly prepared governments scrambled to respond in real-time as the situation developed without the benefit of a well-rehearsed playbook.

Covid-19 is the name of a disease (Coronavirus Infectious Disease-2019) caused by a virus called SARS-CoV-2, meaning it is version 2 of the SARS (Severe Acute Respiratory Syndrome) coronavirus. In the past, pandemics usually caused chaos and confusion because they were something new and came out of the blue, but this wasn't even a new virus; in fact, it was the third coronavirus outbreak we know of. New outbreaks now seem to be happening roughly every ten years. The first version of SARS was identified in humans in 2002 and was twenty times more lethal per infection than Covid-19. Another outbreak occurred in 2012 when Middle East Respiratory Syndrome (MERS) was found in the Middle East. SARS began in a rural area of China and later spread to around thirty countries, causing 774 deaths from just over 8,000 reported cases between 2002 and 2004. Until then, coronaviruses were only known to occur in animals. In May 2003, a study was carried out that tested wild animals sold as food in a market in Guangdong, China, and found a virus very close to the SARS coronavirus in the masked palm civet, a small nocturnal animal that looks like a cross between a cat and a mongoose. Scientists identified several similar viruses in Chinese bats two years later, proving they had jumped across species. MERS also spreads between animals and people, this time with camels being the likely source of the virus. After a fourteen-year-long search, Chinese scientists eventually traced the SARS virus to a specific colony of cave-dwelling bats from Yunnan Province in China. Given current evidence, the most likely explanation for the origin of Covid-19 is that it spread from bats to humans in Wuhan sometime in 2019. While both SARS and MERS are like Covid-19 in terms of origin and virology, the impact and spread of the outbreak was very different. Covid-19 might have been less deadly than its predecessors, but it was much more transmissible and spread so fast it took authorities by surprise. Given the region's ability to contain previous SARS outbreaks, why wasn't it stopped in its tracks at the very beginning?

The city of Wuhan sits almost in the very centre of China on the mighty Yangtze River and is the sprawling capital of Hubei province with a population of eleven million people. It is in this busy commercial interchange that Covid-19 first surfaced, and the first lockdown took place in January 2020 in a belated attempt to contain it. In late December 2019, China informed the WHO that it was treating dozens of patients in Wuhan with pneumonia of an 'unknown cause'. A few weeks later, staff in the central hospital in Wuhan were banned from talking about the disease, and reporters from Hong Kong were arrested and taken to a police station after trying to film what was going on inside the hospital. No one knew how serious this would become, and people were still travelling in and out of Wuhan for a month after the initial outbreak. The first confirmed case outside China occurred in the US when a man in his thirties developed symptoms after returning from a trip to Wuhan. Then, on Thursday, 23rd January 2020, Chinese authorities banned all transport to and from the city, suspending buses, the subway and ferries and closing airports and train stations. The emergency lockdown caused panic, as residents emptied supermarket shelves, hoarded supplies and isolated themselves at home. Petrol stations ran out of fuel and pharmacies quickly sold out of face masks. By the time the city was sealed, over 100,000 people had already fled the city by train to escape infection. This sudden lockdown came almost a month after the first cases appeared, and people were angry with the Chinese government, which had spent the previous few weeks telling the public that the virus was not serious and was 'controllable'. In these crucial first few weeks, information about the emerging virus was kept secret or downplayed, with China's Centre for Disease Control insisting that it wasn't being transmitted from human to human but had come from a 'wet market' or wildlife market in Wuhan. While health experts and doctors around the world started to pick up on signs of alarm, China refused to allow a WHO team into the country to help and began to carefully control all communications about the crisis. On 30th January, the WHO declared a global health emergency after thousands of new cases in China.

With such a highly infectious virus, early detection and warning are vital if it is to be quickly contained. Early warning also helps other countries prepare for it and take precautions to limit the spread, kickstarting the fightback by international scientists and health experts. We will never know if Covid-19 could have been contained and the scale of the tragedy reduced, but China's failure to alert the international community as quickly as possible, share information openly and allow overseas experts to participate cost almost four weeks in a situation where every second counts. Ian Jones, professor of biomedical sciences at the UK's Reading University said in April 2020, 'There was an early cover up in Wuhan, perhaps a few days to a week, before the threat was accepted. We will never know if faster action in those first days could have averted the outbreak.'[5] Information was either ignored or suppressed, and anyone who tried to raise the alarm was severely punished. Chinese whistle-blower Dr Li Wenliang was the first to try to warn people outside Wuhan, sending a message on 30th December 2019 to his medical school alumni WeChat group, warning them that several patients from a local seafood market had been put into isolation after being diagnosed with SARS. Dr Li was called in by police and reprimanded for spreading rumours and disrupting social order. On 28th January, China's top court criticized the police for its actions during the early days of the outbreak, and this was welcomed by Dr Li, who told the media: 'I think there should be more than one voice in a healthy society, and I don't approve of using public power for excessive interference.'

One week later, he died from the disease after treating an infected patient. He was thirty-four years old.

The blame lies with government officials burdened by China's slow, defensive and centralized bureaucracy. In contrast, medics and scientists in China were eager to share information with their colleagues around the globe and played a vital role in understanding the new virus. As the pandemic exploded around them, a team led by Zong-Zhen Zhang, of the Shanghai Public Health Clinical Centre and School of Public Health, raced to sequence the new virus, publishing the initial viral genome on two open-access sites within a few weeks. It was the fastest sequencing effort ever and was quickly followed by the publication of the first proper description of the disease by Chinese doctors writing in the *Lancet* medical journal. During the first month of 2020, most countries, including America, Europe, and the UK, believed the threat was low. In the UK, the government held an emergency COBRA (Cabinet Office Briefing Room A) meeting on 24th January chaired by Health Secretary Matt Hancock, during which England's chief medical officer, Professor Chris Whitty, said the risk to the UK public was 'low'. On 28th January, the same day that Dr Li was cleared, the *Wall Street Journal* ran an editorial entitled 'Act Now To Prevent An American Epidemic', written by two of the federal government's chief scientists. They warned that the virus was spreading fast and pressed for mass testing and travel bans. The White House responded by creating the Coronavirus Task Force and two days later banned entry into the United States by anyone who had visited China in the previous fourteen days. Over the next month, more and more cases started appearing around the world despite travel restrictions. It slowly became clear this would be a global emergency, yet many world leaders underestimated the true scale of what was to come. During a press conference on 25th February 2020, US President Donald Trump said, 'China is working very, very hard. I have spoken to President Xi, and they're working very hard. And if you know anything about him, I think he'll be in pretty good shape. They've had a rough patch, and I think right now they have it – it looks like they're getting it under control more and more.

They're getting it more and more under control. So, I think that's a problem that's going to go away.' By 11th March, 118,000 cases had been reported from 114 countries, and the WHO Director-General Tedros Adhanom Ghebreyesus declared Covid-19 a global pandemic, calling on all countries to take 'urgent and aggressive action'.

The WHO encouraged countries to prepare fast and highlighted the need to 'detect, protect and treat'. They recommended a response that consisted of strict personal hygiene measures, contact tracing and isolating when ill. While the need to detect the virus involved the application of new technology, other measures hadn't changed in hundreds of years: closing borders, quarantining the sick or encouraging self-isolation, face masks, and washing your hands. The things most people could do to protect themselves were very familiar. The ancient Greeks used quarantine to prevent the spread of infectious disease and, as the plague spread along the ports of the Mediterranean during the fourteenth century, city-states posted armed guards along entry points to control entry. In 1821, the Duke of Richelieu deployed French troops to the border between France and Spain to prevent yellow fever from spreading into France, instituting the first *cordon sanitaire*. As the epicentre of the outbreak, China was the first country to shut down, sealing off Wuhan and then the rest of the country in a brutal national lockdown. Other countries in East Asia, such as Vietnam, did the same. Italy became the first European country to introduce a national lockdown, as did Spain and France shortly afterwards. The pandemic resulted in the largest number of national lockdowns at the same time in history. By 26th March, 1.7 billion people worldwide were in some sort of lockdown, rising to 3.9 billion by the beginning of April – more than half the world's population. Countries like Taiwan, South Korea and Singapore, which had experienced previous outbreaks of contagious diseases like SARS in 2002–3, had more success controlling Covid-19, at least in the first few months. They were better prepared and already had mechanisms in place to detect, trace and control disease outbreaks. In these countries, the public was familiar and more compliant with the restrictions needed to prevent the spread of infection.

Governments around the world told citizens to 'stay home, wear a mask, and wash your hands' – much the same advice they had given during Spanish flu a hundred years earlier. Shops and pharmacies sold out of face masks, people hoarded what masks they could find, and a global mask shortage ensued. As Covid-19 mushroomed, people worldwide scrambled for masks to try to fend off the rapidly spreading pandemic. 'The surgical face mask has become a symbol of our times', wrote the *New York Times* on 17th March 2020. The protective qualities of masks are well known, with surgical masks being an important tool used by doctors and nurses everywhere to prevent them from sharing infection with their patients. In Europe, during the Middle Ages and Renaissance, doctors treating plague victims would wear black cloaks, dark hats and masks with a long beak-like nose, sometimes filled with herbs such as clove and cinnamon, to protect them from the poisonous vapour that they believed carried disease. These doctors became popularly known as 'Beak Doctors', an emblem of the darkness of the plague years. Mask-wearing has been commonplace in Asian countries for many years either to protect against choking city smog or as a legacy of previous epidemics in the region. China was one of the first countries to make mask-wearing in public mandatory as part of their strict efforts to control the spread. Drones were used to admonish people seen in public not wearing their masks, and videos of these drone warnings became a social media hit on the Chinese Weibo platform. One elderly resident of a village in Inner Mongolia was startled when she heard a disembodied voice from a drone tell her: 'Yes, Auntie, this is the drone speaking to you – you shouldn't walk about without wearing a mask!' Just like in the United States during the Spanish flu, 'mask slackers' were publicly shamed. By August, over fifty countries had made mask-wearing mandatory. In Austria, Chancellor Sebastian Kurz acknowledged that wearing them would require a 'big adjustment' because 'masks are alien to our country'. Later, mask-wearing became a political issue with protests in the United States, where some saw mask laws as an attack on personal freedom, and in the UK, where protesters raised concerns about the effectiveness of masks as an anti-Covid measure. A survey of over 2,000 adults from the US and Canada carried out between July and August 2020 showed that 84 per cent of people wore masks because of Covid-19. The 16 per cent who did not wear masks refused because they had a negative attitude towards them. They found this was mainly due to one of two things: a belief that masks were not helpful in protecting you from Covid-19 or 'psychological reactance', a state of mind that is aroused when people experience a threat to the loss of free behaviour. In other words, people don't like to be *forced* to wear masks.[6]

Copper engraving of Doctor Schnabel (Dr Beak), a plague doctor in seventeenth-century Rome (Paulus Fürst, 1656)

Most citizens obeyed restrictions and did what they could to follow guidance issued by governments and public health authorities. They stayed at home, only venturing out for essential business, isolated from vulnerable loved ones, wore masks and washed their hands or used hand sanitizer everywhere they went. Offices were closed, and workers were told to work from home. Tourism and travel ceased. Cities across the world were eerily deserted: Times Square in New York, the City of London and the Place de la Concorde in Paris became ghost towns in what the *New York Times* described as 'the Great Empty'. During the first wave of Covid-19, lockdowns lasted until at least June in most countries before restrictions started to be eased. Europe was the first region outside of China to experience a surge, which peaked in April with the highest number of deaths in Italy, perhaps due to it having a higher proportion of elderly citizens than any other country. As the number of cases began to ease off in Europe, they rapidly increased in the Americas and by the end of May, 100,000 people in the US had died. Nine months after the first cases appeared in China, on 29th September 2020, the number of people who had died from Covid-19 reached one million. One fifth of those deaths happened in the United States.

As early as April 2020, over ninety vaccines were being developed by research teams and universities worldwide in a global race to perfect a vaccine and manufacture it in sufficient quantities to combat a planet-wide pandemic. On 8th December 2020, May Parsons, a matron at University Hospital Coventry in the UK, became the first person in the world to administer a Covid-19 vaccination outside of a clinical trial. Her patient was a ninety-one-year-old woman called Margaret Keenan, who was desperate to get back to being able to see her family again, especially her grandchildren. In the first year after they were introduced, vaccinations prevented an estimated 19.8 million deaths from Covid-19 in 185 countries, a reduction of about 63 per cent.[7] Another wave of the pandemic occurred in January 2021 as new variants appeared, and at its peak, the global death rate reached more than 100,000 people per week. As new vaccines became available and were distributed worldwide, healthcare workers raced to vaccinate as many people as possible as quickly as possible. By 2022, thirteen billion doses had been administered globally. In May 2023, after the number of deaths had plunged by 95 per cent, the WHO declared an end to the international health emergency, saying Covid-19 was here to stay but no longer represented an emergency. Over the three years since it began, Covid-19 had killed more than seven million people worldwide and caused the worst global recession since the Great Depression in 1929.

While the biggest achievement of the Covid-19 pandemic was the incredible speed of development and deployment of a vaccine, the biggest failure was how unprepared many countries were and how badly political leaders managed the response. The most fundamental role of government is to keep people safe, especially during a crisis, and many people around the world felt they were let down, especially in the UK and United States, where trust in government and politicians hit an all-time low. The 2019 Global Health Security Index – an assessment of 195 countries' ability to respond to a pandemic – was published in October 2019, just months before Covid-19 began. It scored the United States as the best in the world at 83.5 out of 100. Countries such as New Zealand and Singapore, among the most successful in responding to Covid, were not even in the top ten.

Although this index has been widely mocked, it was clear that the United States entered the pandemic with more capabilities than any other country. Yet it was hit harder than almost any other country, had one of the highest death rates and is ranked in the top five for overall number of Covid-19 deaths. So what went wrong? Despite its score for pandemic preparedness, it was slow to develop and distribute tests, struggled to trace contact between infected people and failed to adequately isolate and quarantine those exposed. One of the biggest mistakes was the way in which Donald Trump's government downplayed the danger and ignored experts. Speaking to the press from the White House Rose Garden on 24th March 2020, Trump showed his complacency: 'We lose thousands and thousands of people a year to the flu. We don't turn the country off. And actually, this year we're having a bad flu season. But we lose thousands of people a year to the flu. We never turn the country off. We lose much more than that to automobile accidents… I would love to have the country opened up and just raring to go by Easter.' During the crucial first six weeks of the crisis, Trump continually blamed China for what he called the 'China virus', told reporters it was under control and insisted 'it will go away'. The administration also carefully controlled what the US Center for Disease Control (CDC) said to the public and restricted top government health experts from communicating accurate and life-saving information.

During the early stages of the pandemic, researchers at the Blavatnik School of Government at the University of Oxford studied how different countries responded in the early days of the outbreak. They looked at how strict the government measures were and included data such as school closures, event cancellations, public information campaigns and emergency investment in healthcare. They created a Stringency Index for over seventy-three countries and tracked it over time, seeing how quickly government policies escalated as the number of cases increased. It showed that the USA was extremely slow to introduce new measures and lagged behind almost every other country. The Stringency Index for the United States didn't change for forty days after the first reported case, and even then, it remained lower than in other countries. America wasn't alone in floundering during the onset of the pandemic: the UK also performed 'significantly worse' than many other countries. A joint report from the House of Commons Science and Health Committees published in October 2021 made it clear that Britain's failure to impose a lockdown early in the Covid-19 pandemic cost 'thousands of unnecessary deaths' and 'ranks as one of the most important public health failures the United Kingdom has ever experienced'.

An Ipsos poll in September 2020 found that most Americans believed President Trump moved too slowly, and around two thirds (68 per cent) didn't trust what he said about the pandemic.[8] During the Spanish flu pandemic, information and advice from government leaders and health experts was clear, consistent and unambiguous. Not only did the public understand what they could do to keep themselves safe, but they also had confidence in the information they were being given and were happy to obey any restrictions. While the things people could do to keep themselves safe during Covid-19 was much the same as in 1918, there was clearly a big difference in the clarity of the information being communicated and how much people felt they could trust their leaders. Covid-19 was the first pandemic of the internet age, when fake news or conspiracy theories spread like wildfire and gained traction with people who doubted the information being presented through government channels and traditional media. As early as mid-March, around half of Americans (48 per cent) said they had seen at least some information about Covid-19 that seemed 'completely made up'.[9] In this case, misinformation wasn't limited to social media, but was also being communicated by political leaders like President Trump. In one Covid-19 press conference, Trump said they were investigating how sunlight could fight the virus and seemed to suggest that injecting disinfectant might be a cure, causing a widespread outcry from medics worried people might actually try drinking bleach. While most wrote it off as simply the latest in a series of bizarre and surreal presidential musings, accidental poisonings from disinfectant rose 121 per cent in the month after the press conference.

In general, trust in scientists increased during Covid-19. According to the Wellcome Global Monitor – the largest survey of how people think and feel about science – levels of trust in scientists worldwide rose significantly between 2018 and 2021. The situation was more complex in the US, where political interference and oversight of the CDC and government health experts by the White House often prevented health officials from saying what they felt was right and made people suspicious of their advice. A CDC official told CNN, 'We've been muzzled. What's tough is that if we would have acted earlier on what we knew and recommended, we would have saved lives and money.' The politicization of Covid-19 communication undermined public confidence in the very experts who could provide the best information and hampered their ability to mount an impartial and coordinated response. A poll carried out by online news magazine *The Conversation* asked members of the public in the UK, US, Germany and Italy how much of the time they trusted their government during Covid-19. They used a four-point scale, with 0 meaning 'just about never' and 3 being 'just about always'. The US came out with a low average score of 0.85, compared to quite a lot of trust in Germany at 1.65. The UK was 1.1 and Italy 1.28.[10] Low levels of trust and widespread disinformation had serious consequences – amplifying public fears and leading to harmful behaviour. Research shows that trust in authorities has a big influence on the actions people take, especially during a crisis. An analysis of US survey data collected between 27th and 30th March 2020 found that people who trusted health experts were more likely to take the threat from Covid-19 seriously and take protective measures such as wearing a mask. The same study also showed that people who trusted the White House leadership thought themselves safe and were less likely to take protective actions.[11] Possibly one of the most serious effects of this situation was the incredibly high number of people who refused to get vaccinated, even at the height of the pandemic. According to the Kaiser Family Foundation, 230,000 deaths could have been avoided if individuals had got vaccinated earlier.

One of the best lessons we can learn from Covid-19 is the vital importance of clear, transparent and reliable information. The way this is communicated to the public and how much trust the public have in both governments and experts is critical during a global emergency in helping people respond promptly and take the right actions to keep themselves safe. A team led by Professor Jia Liu from the University of Portsmouth has studied the global data from the International Coronavirus Survey and found that during public health emergencies, governments must be accountable, act quickly and establish frank and timely dialogue with the public to encourage trust and cooperation and alleviate fear. After analysing 111,196 responses from 178 countries in March and April 2020, Professor Liu 'found that individuals are positively influenced by the fairness, effectiveness and accountability of government agencies, plus public information campaigns. Honest communications keep citizens informed, help them to understand the pandemic, prevent scepticism and strengthen trust in government.'[12]

Disease X

Ebola, Marburg virus disease, Lassa fever, Rift Valley fever, Zika and Crimean-Congo haemorrhagic fever are all on the WHO priority list for diseases that have the potential to become a serious future epidemic. Most of these originated in animals and are zoonotic diseases transmitted from animals to humans. In fact, 75 per cent of all new diseases are zoonotic. Acknowledging the unknown, they also list 'Disease X' – a serious international epidemic that could be caused by a currently unknown pathogen. Who had heard of SARS-CoV-2 in 2019? With an estimated 1.6 million viruses lurking in the world's mammalian and avian wildlife, whether next year or in ten years, Disease X is inevitably on its way. Like previous pandemics, it will be a hyper-virulent disease that no one has seen before, with no vaccine or treatment we know of, and could kill hundreds of millions of people. Chief investigator for the clinical trials of the Oxford Covid-19 vaccine, Professor Andrew Pollard, told *The Times*: 'We can't take our eyes off the ball because globally we are in a perfect situation for future pandemics. We should expect these threats to become more common, thanks to the fact that the world is getting more crowded, people are travelling internationally more, and populations are encroaching more on previously wild environments. It's a perfect melting pot.' Planetary trends are driving an increase in the likelihood of zoonotic spillover. Habitat destruction and increased urbanization are putting humans and animals in closer contact, climate change is forcing animals to migrate to new territories, causing many species to meet for the first time, and travel and supply chains are making the world more interconnected than ever before. So, how prepared are we for the next pandemic? In the aftermath of Covid-19, we are thankfully more prepared than ever before – it was a wake-up call to every country that pandemic preparedness is a priority and needs serious investment.

Scientists are already working on vaccines for diseases on the WHO priority list. Professor Sarah Gilbert, one of the co-creators of the Oxford Covid vaccine, is back working on the MERS coronavirus and Nipah virus in the hope of deploying a vaccine before an outbreak occurs. Due to existing research, we had a head start with the Covid-19 vaccine, which meant it could be created very quickly and deployed on a massive scale thanks to streamlined clinical trials and successful collaboration between researchers and pharmaceutical companies. Many countries have now committed to the '100-Day Mission' – investing in the ability to create a safe and effective vaccine and get it in people's arms within 100 days of sequencing the genome of a new virus. There is a renewed global effort to improve disease surveillance and monitoring so that new outbreaks can be identified quickly and information shared openly. Serious investments are being made in the infrastructure needed to cope with future pandemics. The World Bank has launched a Pandemic Fund to provide critical funding to low- and middle-income countries to help them build their capacity to fight future outbreaks. And in September 2023, the first-ever heads of state summit on pandemic prevention, preparedness and response was held during the UN General Assembly. Member states committed to working together to strengthen international cooperation and investment to prevent a repeat of the devastating impact of Covid-19, and we are close to finalizing a global pandemic treaty. Responding to the summit, WHO Director-General Tedros Ghebreyesus said:

We must learn how to protect our communities better and to engage, inform and empower them to be part of the solution. We need to build stronger clinical care systems that can save lives. Doing so requires concrete actions to ensure equitable access to medical countermeasures, sustainable and adequate financing, empowered and engaged communities and robust, trained and equipped health workers.

Pandemics cross borders, so international collaboration is vital – global problems need global solutions.

The Dutch humanist scholar Erasmus, whose parents died from bubonic plague in 1483, was one of the first to point out that 'prevention is better than cure'. Instead of preparing for a future pandemic, we should do everything we can to prevent one from happening. Monitoring for new pathogens in animal populations, better regulation of wildlife trade and addressing deforestation could dramatically reduce the likelihood of zoonomic spillover. Between 2000 and 2012, 2.3 million square kilometres of forest were lost, making it the greatest threat to biodiversity and a leading cause of the emergence of new diseases. If we can't prevent it, then we must at least make sure we are more resilient, but what would a pandemic-resilient world look like? In the future, new outbreaks would be identified within hours of the first few cases occurring, and instant alerts would be sent to the global pandemic community on day one. Containment protocols would be automatically triggered to try to stop it in its tracks before it becomes a pandemic. But, if it did spread, local communities all around the globe would have central hubs containing stockpiles of food, masks, tests and other essential supplies that they would distribute in emergency survival packs to all residents by drone delivery. A rapid-response force would initiate a vaccine programme that deployed vaccines worldwide within 100 days. Every member of the public would already carry a digital immunity passport that tracks their exposure and immunity and advise on travel and self-isolation. Vaccines stop diseases, but so does money. Special funding would immediately be released to countries that need it for preprogrammed response efforts. Businesses would seamlessly switch to home working to protect their staff, and national authorities be guided by a 'warbook', based on the best expert advice and research available. It would tell them the optimum time to lock down or close schools and what other measures they should take. But will the political leaders of the future stick to the plan in the heat of a sudden crisis?

Chapter 5

When the Levee Breaks

The year 2005 was characterized by worldwide instability, conflict and terrorism. The war in Iraq was entering its third year and the London bombings on 7th July sent shockwaves through Europe, with most countries raising security levels and a deepening fear of further attacks among the public. It is also remembered as a year of natural disasters, with earthquakes and tsunami waves in Asia, hurricanes in Central and North America, a major earthquake in Pakistan and India, a locust infestation that caused a famine in Niger and a volcanic eruption in tiny El Salvador that displaced over 7,500 people. Hurricane Katrina was a tragedy that devastated coastal Louisiana and its neighbours, particularly the city of New Orleans, in late August 2005, just eight months into George W. Bush's second term as US president. It was the costliest tropical cyclone on record, a deadly Category 5 Atlantic hurricane that killed 1,836 people, caused $145 billion in damages and led to a global spike in oil prices. Japan's biggest-selling newspaper, the *Daily Yomiuri*, cited the hurricane as the most significant international news story of 2005, observing:

> The Bush Administration faced severe criticism over delayed relief operations and dealing with such crimes as looting. African Americans suffered most as they lived in the hardest-hit areas. Some people say this highlighted the country's racial divisions and rich-poor gap.

Merriam-Webster, the oldest dictionary publisher in the US, gives a telling insight into public sentiment in their 'Top Ten Words of the Year for 2005' (based on online searches): integrity, refugee, contempt, filibuster, insipid, tsunami, pandemic, conclave, levee and inept.

The biggest risk to the people who live along the Mississippi Gulf Coast comes from water. Whether caused by snow melting in the north or by intense rainfall or a storm surge from a hurricane – it's the Great Deluge that follows that claims most lives and destroys entire communities. In the case of Hurricane Katrina, most deaths were due to flooding caused by flaws in the flood protection system, especially the levees around the city of New Orleans. Given Louisiana's long history of flooding, what exactly went wrong, why weren't people better prepared, and how can we keep communities safe when the next big hurricane strikes?

Holding Back the Flood

The southern coastline of Louisiana is some of the youngest land in the United States. At the end of the last ice age, roughly 20,000 years ago, the great glacial ice sheets melted, causing sea levels to rise by 120 metres. Coastal areas were submerged, becoming continental shelves, and shorelines retreated before reaching a stable point about 7,000 years ago. Today, most of America's coastal residents live along the same shorelines. However, sitting in the middle of the Gulf of Mexico, part of Louisiana shaped an enclosed bay into which the mighty Mississippi and other great rivers flowed. When the rapid rise in sea level finally slowed, sediment from thirty-one US states and two Canadian provinces was carried down by these mighty waterways, filling up and reclaiming the shallow Gulf of Mexico. It created the enormous deltas, marshlands and swamps that make up Louisiana's Mississippi Delta Gulf Coastline, extending for over 400 km from Vermillion Bay in the west to Chandeleur Sound in the east. With most available land just a few feet above sea level, it's a fragile landscape prone to regular floods, tropical storms and catastrophic weather. Extreme rainfall in the summer of 2022 led to flooding that overwhelmed water treatment plants in Jackson, Mississippi, killing three people and leaving more than 150,000 without drinking water. It's also one of the fastest-disappearing areas in the world. Between 1932 and 2010, the state lost 1,800 square kilometres due to coastal erosion and sea-level rise, and researchers have estimated that thirty football fields' worth of land is being lost every day. It is hard to imagine why anyone would want to live in such a difficult and obviously risky location, but its strategic importance at the mouth of the Mississippi River was evident to the earliest settlers.

The second-longest river in North America, the Mississippi runs 2,340 miles (3,770 km) from its source in the glacial lake of Itasca in northern Minnesota to the Gulf of Mexico. It was 'Misi-ziibi' (the 'Great River') to Native Americans, who have lived along its banks for thousands of years, using it as both highway and larder, travelling from village to village in dugout canoes and fishing as their primary source of food. By 500 AD, they had developed a network of trade routes along the whole river valley, linking communities from the Great Lakes all the way to the Gulf of Mexico. Later, it would emerge as a vital and speedy transportation route carrying furs from Canada and the Great Lakes, corn and wheat from the Midwest, and cotton, sugar and tobacco from the Deep South. While trade flourished, the river was dangerous and unpredictable. Following Columbus's discovery of the New World in 1492, young Spanish nobles were attracted by tales of adventure, glory and wealth – setting out on expeditions to the Americas. The *conquistador* Hernando de Soto was the first European to negotiate the river in the hope of plundering the native tribes. A fine horseman and daring adventurer, de Soto first came to the New World in 1514, exploring central America and amassing considerable wealth. He joined Francisco Pizarro in the conquest of Peru, returning to Spain in 1536 a rich man. Restless and in search of greater fame, he returned in 1539, landing on the west coast of Florida with 600 men, 200 horses and a pack of bloodhounds. They scoured Florida, Georgia, South Carolina and Alabama, subduing the indigenous people and seizing any valuables they found, but the gold and silver they desired eluded them. In May 1541, his raiding party reached the Mississippi just south of what is now Memphis, Tennessee, becoming the first Europeans to cross the river. The Native American tribes gathered together in an attempt to repel the invaders and, after suffering heavy casualties, were given a helping hand by the sudden Mississippi floods that caught the Spanish unaware. De Soto died of a fever on the western bank of the river on 21st May 1542. To preserve the myth that he was a god ('the Immortal Son of the Sun'), his soldiers hid his corpse in blankets, weighed it down with sand and sank it in the Mississippi in the middle of the night. Under heavy fire from the Native Americans, the remaining army built rafts and retreated down the river, finally reaching Mexico in 1543.

The French claimed the territory of Louisiana in 1682 and, after a brief period of Spanish rule, it was acquired again by France in 1801 as part of Napoleon Bonaparte's dream of building a great French empire in the New World stretching from the Mississippi valley to the West Indies and beyond. The French were in control not only of the river, but also of its great port, New Orleans. Afraid of France's colonial ambitions, US officials were worried that it would dominate an important trading port and restrict access to the Gulf of Mexico. One of the Founding Fathers, Thomas Jefferson, noted, 'There is on the globe one single spot, the possessor of which is our natural and habitual enemy. It is New Orleans.' Napoleon soon lost interest in establishing a North American empire and, in dire need of money to fight the British, sold Louisiana's 828 square miles to the US in 1803 for $15 million, instantly doubling the size of the fledgling American republic. Almost one hundred years earlier, the governor of French Louisiana had founded the city of Nouvelle-Orléans (New Orleans) on the first crescent of high ground above the mouth of the Mississippi River, just twelve feet above sea level. The risk of frequent flooding was well known, but it was the only possible location for a port, becoming what geographer Peirce F. Lewis called an 'impossible but inevitable city'. Flooding was immediately a problem, interrupting construction, and the following year, the new settlement was destroyed by a hurricane so strong it caused a surge that reversed the flow of the river. French engineer Adrien de Pauger wrote, 'The river rose more than six feet and the waves were so great that it is a miracle that [all the boats] were not dashed to pieces.' De Pauger was responsible for planning the city's rebuild and designed the grid for the French quarter which is still in use today.

> One who knows the Mississippi will promptly aver... that ten thousand River Commissions, with the mines of the world at their back, cannot tame the lawless stream, cannot curb it or confine it, cannot say to it Go here or Go there, and make it obey; cannot save a shore that it has sentenced.
> – Mark Twain, *Life on the Mississippi* (1883)

Most people's understanding of life along the banks of the Mississippi in the eighteenth and ninenteenth centuries comes from the 'father of American literature', Mark Twain, who grew up just a few miles away and spent four years piloting steamboats on the river before the Civil War. Ernest Hemingway once said 'all modern American literature comes from one book by Mark Twain called Huckleberry Finn'. Inspired by his days on the riverboat, *The Adventures of Huckleberry Finn* chronicles the people, places and politics of the South in the pre-Civil War years. Throughout the book, Twain portrays the river as a powerful force, constantly causing destruction and changing the landscape along its banks, yet a symbol of freedom to the primary characters, Huck and Jim. In the book, it is clear to Huck that the only thing about the Mississippi you can rely upon is its ability to surprise and thwart attempts to tame it. Yet that is exactly what generations of engineers and town planners have attempted to do in doomed efforts to protect communities from the worst impact of flooding. New Orleans itself was built on a natural 'levee', a ridge of sediment deposited along the banks of a river that helps prevent it from overflowing, from the French verb *lever*, meaning 'to raise'. Building up these natural barriers was the first solution attempted by the French in 1717 to defend their infant city. Over the next two hundred years, landowners were forced to build a complex array of embankments in an increasingly desperate attempt to contain or control the waters. The first levees were only a few feet high and woefully insufficient to hold back the deluge during fierce floods in which the river would frequently break its banks, taking lives and damaging property. Later, the Army Corps of Engineers was drafted to improve the design of the levee system, and its increasing effectiveness allowed New Orleans to expand. In some ways, the levees were too good at containing the water, and the better they were, the more they prevented water from escaping over the levees or through natural channels, thus raising the flow of the river and increasing the threat of flooding. Engineers spent years debating the merits of a levee-only strategy and whether they should incorporate some system to help regulate the level of the river, but everything changed following the Great Mississippi Flood of 1927.

At 8 a.m. in the morning of 21st April 1927, just north of Greenville, Mississippi, 13,000 people were stranded on a strip of levee just eight feet wide and five miles long. Surrounded by a raging torrent, they had no water, food or shelter for several days until Red Cross rescue barges could get to them and transport them to safety. Some of the relief barges would only pick up white people, leaving Black families to set up temporary camps along the levee. Similar scenes could be found all across Arkansas, Mississippi and Louisiana. The Great Mississippi Flood was the most destructive flood in the history of the United States and remains one of the greatest environmental disasters ever. It happened early in the spring, as streams in the north were swollen by melting snows, and a low-pressure system hung over the US for many months, bringing with it a sustained period of intense rainfall that saturated the ground. Inadequate levees sagged and snapped in over 145 different places, and the raging waters of the Mississippi broke through, pouring over the land, wreaking havoc, distress, misery and death from Illinois to the Delta. Over several months, 27,000 square miles of land was inundated by flood water up to thirty feet (9 m) deep. In Vicksburg, Mississippi, the river swelled to eighty miles wide. Amazingly, only 500 people lost their lives, but 630,000 were affected. One observer described the flood waters as 'a torrent ten feet deep, the size of Rhode Island, it was thirty-six hours coming and four months going; it was deep enough to drown a man, swift enough to upset a boat, and long enough to cancel a crop year'. Another survivor said the river was 'a tossing, seething yellow sea as far as the eye can reach... houses and familiar objects look grotesque... cut in two by the climbing oblivion of the water line'.[1]

Floods were a normal part of life throughout the south and African Americans, who made up 95 per cent of the agricultural labour force, were always hit hardest. Inspired by a flood the previous year, blues singer Bessie Smith released 'Back Water Blues', in which she sings heartbreakingly:

When it thunders and lightnin',
And the wind begins to blow,
There's thousands of people,
Ain't got no place to go.

More than 200,000 African Americans were displaced from their homes and had to live for long periods in relief camps set up by the American Red Cross. Railroads and plantations affected by the floods feared their labourers, many of whom had lost everything, would move away and never return. At many of the camps, the National Guard forced Black refugees to carry out heavy labour and prevented anyone from leaving, promising to return them to their employers after the crisis was over. One sign in Greenville announced that 'refugee labour is free to all white men', and they were forced to unload food and supplies and repair homes or plantations flattened by the flood. Herbert Hoover, who was then Secretary of Commerce, led the federal government's response and built the world's longest system of levees and floodways to prevent such a disaster from happening again. The Flood Control Act of 1928 shifted policies from levees-only to one which also included spillways and other structures to better control the immense volume of water experienced during these catastrophic events. Hoover was hailed as a national hero for his handling of the crisis, catapulting him into the presidency in 1929, although he was later heavily criticized for covering up the Red Cross's treatment of Black workers who had been kept in 'concentration camps'. More than fifty blues songs were recorded in 1927 about the tragedy and the injustices suffered by people still trapped in the camps. Louisianan Memphis Minnie and Joe McCoy recorded 'When the Levee Breaks' in 1929, when the impact and upheaval caused by the flood were still fresh. Minnie had been living with her family in Walls, Mississippi, when a levee broke in 1927. Later reworked and released by Led Zeppelin, the song recounts the personal toll on a man who lost his home and family when the levee broke at Greenville: 'What should I do? I've got nobody to tell my troubles to.'

Okeh Records advertisement for Blue Belle's
'High Water Blues', May 1927

Since 1928, the US Army Corps of Engineers has invested over $10 billion of taxpayer money to build up the flood defences along the river, making it much easier for a large volume of water to move downstream and out into the Gulf of Mexico without spilling over into fields, highways and homes. Another 'super flood' in 1937 caused significant damage, and the loss of 244 lives, but none of the main branch levees broke. Over the next fifty years, the Corps continued to make improvements in the channels, and it is estimated that their work prevented around $244 billion in flood damages. A flood in 1993 was similar in magnitude to that in 1937, but this time, the river was kept at a much lower level, so no lives were lost, and it caused minimal damage. With better design of levees, floodwalls and floodwater pumping stations, it seemed the engineers were winning the war against nature. But we now know that the thousands of miles of levees and flood barriers that were built to protect local communities have also eroded the natural defences of the Delta, making New Orleans and other towns more vulnerable to storm surges and hurricanes. The wetlands, marshes and backswamps of Louisiana's coastal zone are the first line of defence against storms but are disappearing at a rate of twenty-three square miles every year. This unique young habitat is home to black bears, bottlenose dolphins and beavers, as well as several rare birds, including the indigo bunting and bald eagle. According to the latest research, it is sinking into the Gulf of Mexico at an alarming rate of nine millimetres per year. It has been doomed by the consequences of human action: the need to tame the great Mississippi River, oil-industry canals and sea levels rising faster than ever before due to climate change.

The wetlands used to be rejuvenated each spring by sediment carried by floodwaters, but now the flood protection levees redirect sediment away from the Mississippi Delta, carrying it far offshore. Without this regular replenishment, the delta will eventually sink. Many leading engineers were well aware that levee construction on such a massive scale would have this impact, but put too much faith in the life-saving protection they promised. Elmer Lawrence Corthell, described as 'one of the most prominent engineers in the Western hemisphere', warned that the newly built flood defences would come at a cost. In 1897, he wrote in *National Geographic*:

> No doubt the great benefit to the present and two or three follow-ing generations accruing from a complete system of absolutely protective levees... far outweighs the disadvantages to future generations from the subsidence of the Gulf delta lands below the level of the sea and their gradual abandonment due to this cause. While it would be generally conceded that the present generation should not be selfish, yet it is safe to say that the development of the delta country during the twentieth century by a fully protective levee system, at whatever cost... will be so remarkable that the people of the whole United States can well afford, when the time comes, to build a protective levee against the Gulf waters.

As if that wasn't enough, navigation canals dug for oil and gas exploration have contributed to saltwater intrusion and erosion, furthering land loss and accelerating the subsidence of the delta. All of this, combined with the gradual rise in sea levels in the Gulf of Mexico due to climate change, means that Louisiana is being exposed to the highest rates of relative sea-level rise in the whole of the US. By the time Hurricane Katrina arrived in 2005, New Orleans lay at an average of six feet below sea level.

Global Boiling

Hurricanes are not new to Louisiana or its capital, a high-risk hotspot that sits directly in the path of tropical storms crossing the Atlantic. Most hurricanes start close to Africa and travel westward thousands of miles across the Atlantic Ocean, gathering strength from the warm water as they go. Wind currents direct them towards the Caribbean, Florida and the Gulf of Mexico, although occasionally some drift further north and might curve back out into the Atlantic if the prevailing wind direction changes. Warm waters fuel them, so the peak hurricane season occurs in the hottest months, from August to October. A few years ago, the US National Oceanic and Atmospheric Administration (NOAA) analysed the probability of a tropical storm hitting the US coastline and found that New Orleans had a 40 per cent chance of being hit each year and a 12 per cent chance of it being hurricane strength (with wind speeds of at least seventy-four miles per hour). In *A Furious Sky*, author Eric Jay Dolin tells the story of how hurricanes have shaped the course of US history. Five hundred years ago, a massive storm in the Caribbean sunk twenty-four ships of Christopher Columbus's fleet. England's attempts to colonize America in the early 1600s nearly failed when powerful hurricanes battered some of the first settlements. After a hurricane hit in 1609, the settlement at Jamestown in Virginia almost collapsed, and the colonists were close to starvation when a relief ship finally reached them. An account of the ordeal by one of the survivors gave William Shakespeare the inspiration for *The Tempest*. Further north, the Great Colonial Hurricane of 1635 flattened the English settlements at Plymouth and Massachusetts Bay, destroying houses, sinking ships and taking numerous lives. Resilience in the face of natural disasters was an essential trait for the early American pioneers.

Tropical storms sometimes behave erratically, suddenly changing course or gathering strength overnight, and climate change is making matters worse. Many of the most damaging storms to hit the US in recent years have been remarkable because they intensified so rapidly, surprising forecasters and making it hard for nearby communities to prepare themselves. Hurricane Maria, one of the deadliest and costliest hurricanes to strike the Caribbean, grew from a mere tropical storm to a full-blown Category 5 hurricane in just over forty-eight hours. In 2023, Dr Andra Garner, a climate scientist from Rowan University, investigated Atlantic hurricanes over the past fifty years and found that rapid-onset hurricanes are becoming more likely as greenhouse gas emissions warm the planet and oceans. Garner's research, published in the journal *Scientific Reports*, showed that hurricanes in the Atlantic Ocean are now twice as likely to grow from weak storms into hurricanes within just twenty-four hours. You could go to bed at 10 p.m. with a mild storm on the horizon and wake up to a catastrophic hurricane right on top of you – it's the nightmare scenario for civic authorities, who need time to organize an evacuation and put emergency services on alert. Hurricanes draw their power from the ocean's warmth, so the hotter the surface is, the more thermodynamic energy they can gather. Since 1850, the global average sea surface temperature has risen by about 0.9 degrees Celsius and 2023 saw a record high. A recent statement from the Woods Hole Oceanographic Institution warns that:

> Rising ocean temperatures are making some extreme weather events worse by supercharging storms and altering global weather patterns. Warm surface waters provide energy for hurricanes and other tropical cyclones, increasing their frequency and severity.

Severe hurricanes are getting stronger, becoming more frequent, lasting longer and bringing more intense rainfall. Another study, by the Niels Bohr Institute at the University of Copenhagen, found that the proportion of major hurricanes in the Atlantic Ocean has doubled in the last forty years.[2] For towns in high-risks areas like the Louisiana coast, early warning systems, hurricane defence, and emergency services urgently need to keep ahead of this fast-moving trend.

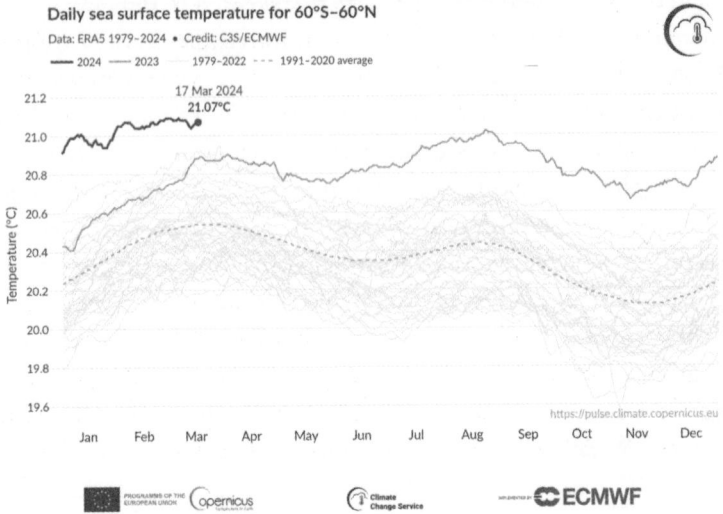

Daily sea surface temperature for 60°S–60°N

Data: ERA5 1979–2024 • Credit: C3S/ECMWF

— 2024 — 2023   1979–2022   - - - 1991–2020 average

17 Mar 2024
21.07°C

https://pulse.climate.copernicus.eu

PROGRAMME OF THE EUROPEAN UNION   Copernicus   Climate Change Service   IMPLEMENTED BY ECMWF

Unprecedented sea surface temperatures are being recorded across the global ocean, with historical record temperatures being measured in vast areas of the tropical seas (European Union, Copernicus Climate Change Service Data, 21st March 2024)

According to the NOAA, since 1851 the top three states for hurricane landfalls have been Florida, Texas and Louisiana. The worst hurricane to hit Louisiana was the Great October Storm of 1893 which struck land at Cheniere Caminada, just west of Grand Isle. Two thousand people were killed by the coastal flood it caused, and not much remains of the once thriving multi-ethnic fishing community that used to supply seafood to the bustling metropolis of New Orleans. Despite the large death toll, it wasn't as strong as Hurricane Betsy in 1965, nicknamed 'Billion Dollar Betsy' because it caused approximately $1 billion in damages. After leaving the Caribbean, Betsy grew in power to become a Category 4 storm, with wind speeds in excess of 130 miles per hour, reaching land at Grand Isle just west of the mouth of the Mississippi. It travelled upriver, causing mayhem and a storm surge in the river of over ten feet, breaking the levees at New Orleans and inundating several neighbourhoods, including the Lower Ninth Ward that lies below an industrial canal, one of the areas hit hardest by Hurricane Katrina forty years later. Affecting many of the same areas, Betsy gave a foretaste of Hurricane Katrina but on a much smaller scale. It killed fifty-seven people in New Orleans, and the tragedy moved President Lyndon Johnson to visit and meet victims of the storm. A new federal Hurricane Protection Programme was put in place to make the levees around New Orleans taller and stronger, designed to resist a fast-moving hurricane like Betsy. But, when Hurricane Katrina smashed through the levees on 29th August 2005, it revealed the design to be faulty and poorly constructed. After nearly 300 years of perfecting flood defences and preparing for future disasters, why did Hurricane Katrina blindside locals and cause so much destruction?

'Certain Death'

The summer of 2005 was one of the hottest ever experienced worldwide. From July, a severe heatwave gripped the south-western United States, and sea surface temperatures in the Atlantic and Caribbean reached a record high. Together with prevailing wind and pressure systems, the global conditions were ideal for storm formation, and so 2005 became the most active hurricane season on record (until it was surpassed in 2020). The first major hurricane of the year was Hurricane Emily, a powerful early season hurricane that formed on 6th July and struck landfall at Grenada in the eastern Caribbean Sea on 14th July, causing extensive damage. Grenada was still recovering from Hurricane Ivan, which had ripped through the islands the previous September, destroying 90 per cent of houses and flattening the Grenada Red Cross headquarters. In the popular tourist destination St George's, Emily blew away the roof of the operating theatre and other wings at the main hospital, as well as the roofs of two police stations in Grenville. There were three other hurricanes that year before Katrina struck: Hurricane Cindy, Hurricane Dennis and Hurricane Irene.

On 8th August, a tropical wave emerged in the Atlantic Ocean from the coast of West Africa. As it travelled westwards, it interacted with other waves, organising into a tropical depression that grew in strength as it moved across the Atlantic. By the time it reached the Bahamas, it was designated Tropical Depression 12 and started to strengthen quickly, turning into Tropical Storm Katrina and heading towards Florida on 24th August. The next day, it reached hurricane strength and, within two hours, made landfall between Miami and Fort Lauderdale, with wind speeds of around eighty miles per hour (130 kilometres per hour). The wind, together with heavy rains, downed trees and power lines, leaving 1.5 million people without power and creating floods that destroyed crops throughout the state. It spent only eight hours over land, gradually weakening, and when it entered the Gulf of Mexico it was mild by hurricane standards. But it was turbocharged by the unusually warm waters of the Florida Loop Current (a warm ocean current that flows between Cuba, the Gulf of Mexico and the Florida Straits), causing it to intensify rapidly and transform from a Category 3 to the strongest Category 5 hurricane in just nine hours. It reached peak strength on 28th August, by which time it had doubled in size and had a wind speed of 175 miles per hour (280 kilometres per hour) – at the time the strongest hurricane ever recorded in the Gulf of Mexico. Its second landfall came the next day, hitting the Louisiana coastline at Plaquemines Parish, forty-five miles south-east of New Orleans, before it continued its course, maintaining Category 3 hurricane strength well into Mississippi. Later that morning, it crossed land again near the mouth of the Pearl River, Mississippi, and caused a giant storm surge twenty-six feet (eight metres) high that smashed into the cities of Gulfport and Biloxi, devastating homes along the beachfront. It gradually dissipated and was absorbed by a cold front when it reached the Great Lakes region on 31st August.

On Friday, 26th August, just after Katrina hit Florida, the National Hurricane Center changed their prediction of its probable path to one which took it over the Mississippi/Alabama coast, showing that it would pass very close to New Orleans, prompting Governor Kathleen Marie Blanco to declare a state of emergency in Louisiana. The National Guard was mobilized and emergency response plans were activated, with the US federal agency FEMA coordinating. But, as Katrina 'took aim' on New Orleans, the response by national officials was shockingly lax. After being briefed by the National Hurricane Centre, FEMA Director Michael D. Brown decided to wait and not immediately send in teams to manage the disaster. An anonymous FEMA team leader who spoke to the *Washington Post* described a strange sense of inaction: 'Nobody's turning the key to start the engine, why aren't we treating this as a bigger emergency? Why aren't we doing anything?' The next day, Katrina was a Category 3 hurricane, and it was clear it would hit New Orleans as a Category 4, or even the deadliest Category 5. Max Mayfield, the head of the National Hurricane Center, told FEMA's daily video conference, 'This one is different. It's strong but it's also much, much larger.' The Center forecast the storm surge could be as high as twenty-five feet, highlighting the possibility that it could overwhelm the levees that protect New Orleans.

While federal agencies hesitated, by Saturday, 27th August, it had become obvious to many local officials this was no ordinary storm and community leaders across the state began to try to get people out of harm's way. Coastal parishes in the direct path of the storm organized a rapid evacuation. Governor Blanco put in place a traffic flow system to help people get to higher ground more quickly and encouraged people in New Orleans to go door to door and persuade their neighbours to leave. New Orleans mayor Ray Nagin had announced a city-wide emergency but held back on ordering the first mandatory evacuation in New Orleans history, the only thing that would have had real force with the public. Worried that a sudden evacuation would cause gridlock, he wanted to give residents of the low-lying parts of the city, such as Algiers and the Lower Ninth Ward, time to leave first. Later the same day, in a joint press conference with Governor Blanco, he urged all residents to evacuate, saying, 'This is not a test. This is the real deal.' But there was no mandatory evacuation order in the city for another twenty-four hours and still people refused to leave their homes.

The next morning, Katrina had escalated into a full-blown Category 5 hurricane, with wind speeds of more than 175 miles per hour, and was fast approaching New Orleans. At 10.11 a.m. on Sunday, 28th August, the National Weather Service issued its most dire warning to date, composed by meteorologist Robert Ricks, who was on duty that day at the Slidell, Louisiana, office that was responsible for New Orleans. Ricks, who had grown up in the Lower Ninth Ward of New Orleans, knew he had to create a stark advisory that would leave people in no doubt of the need to get out and fast:

A MOST POWERFUL HURRICANE WITH UNPRECEDENTED STRENGTH... MOST OF THE AREA WILL BE UNINHABITABLE FOR WEEKS... HIGH RISE OFFICE AND APARTMENT BUILDINGS WILL SWAY DANGEROUSLY... A FEW TO THE POINT OF TOTAL COLLAPSE. ALL WINDOWS WILL BLOW OUT. AIRBORNE DEBRIS WILL BE WIDESPREAD... AND MAY INCLUDE HEAVY ITEMS SUCH AS HOUSEHOLD APPLIANCES AND EVEN LIGHT VEHICLES. PERSONS... PETS... AND LIVESTOCK EXPOSED TO THE WINDS WILL FACE CERTAIN DEATH IF STRUCK.

The Louisiana Superdome, a giant sports stadium in the heart of New Orleans and home of American football team the New Orleans Saints, was opened as a shelter of last resort for people who couldn't get out of the city. By 10 a.m. the next morning, the queue to get in was a quarter of a mile long. When the doors were opened, around 14,000 people streamed inside carrying ice chests, children's toys, clothes and whatever belongings they could carry. They waited as the skies darkened, the air became heavy and humid, and the winds picked up – roaring across the streets, shaking trees and buildings. Over 100,000 people were still stranded in the city, unable to escape before the storm hit. They took refuge in one of the ten emergency centres that had been set up or waited it out at home, praying for a miracle.

Surrounded by vast bodies of water and mostly lying below sea level, New Orleans is a 'soup bowl' just waiting to be filled. Its only protection from being inundated with water was the complex system of levees, pumps and canals built up over the past three hundred years. Early on Monday morning, the hurricane made landfall. While New Orleans avoided a direct hit, it experienced its worst-case scenario: catastrophic rainfall, winds and powerful storm surges that overwhelmed its defences. Levees along the 5.5 km Industrial Canal that cuts south into the city from Lake Pontchartrain were the first to fail, destroyed by a nine-foot storm surge that rolled through the canal. This was quickly followed by more levee breaches nearby, sending water rushing towards the Lower Ninth Ward. Even where the flooding was slower, water rose roughly one foot every ten minutes. Some neighbourhoods flooded to the rooftops in minutes. In total, levees and flood walls failed in more than fifty locations, allowing water from the Gulf of Mexico, Lake Borgne and Lake Pontchartrain to flow into New Orleans. The pump stations were so overwhelmed with water they stopped working, so there was no way to stop the water from rising. By 1st September, over 80 per cent of the city was flooded, much of it six to ten feet (2–3 m) deep. The US Coast Guard helicopter rescue teams carried axes to break through roofs and save people trapped in their attics. Flying over flooded homes, they found whole families waving at them from the remaining parts of their homes still above water. Some had painted signs on their roof in white paint that read simply 'Help Us.'

Many drowned in the sudden onslaught of water, trapped in their homes or swept away while trying to get to a safe place. Some died trying to save family and friends. Others fled to their attics and then had to hack through the roof to find a place to stand in the hope of being rescued. The elderly and sick succumbed as they struggled to get away from the rapidly rising water, and some made it to safety only to die later while they waited for food, water or medical care. Chris Robinson, a resident still at home just east of the downtown area, told the Associated Press by mobile phone: 'I'm not doing too good right now. I got a hammer and a crowbar, but I'm holding off on breaking through the roof until the last minute. Tell someone to get me out please. I want to live.' Water was undrinkable and food scarce. Law and order broke down, and thugs preyed upon the weak. When looting started, FEMA refused to send in emergency workers until the National Guard could provide security, further delaying relief efforts. CNN anchor Jack Cafferty declared:

> I remember the riots in Watts. I remember the earthquake in San Francisco. I remember a lot of things. I have never seen anything as badly handled as this situation in New Orleans. Where the hell is the water for these people? Why can't sandwiches be dropped to those people in that Superdome down there? It's a disgrace. And don't think the world isn't watching.

Power failures hampered rescue efforts; the rescue centres were destroyed, and communication systems broke down, so emergency workers didn't know where to go or who to help. Like the Great Mississippi Flood of 1927, many were stranded for days. Deadly chemicals spilled into the flood water from industrial sites, shops and garages, creating public health problems that compounded the crisis. Overcrowded and flooded hospitals tried to treat people in unsanitary conditions as best they could. Terry Taylor, a teenager who had been wandering the streets for twenty-four hours looking for shelter, told reporters, 'There were dead people floating everywhere you looked.' No one seemed to be in charge, local and federal agencies were paralysed and residents felt abandoned. Despite the chaos, brave emergency workers and volunteers managed to rescue tens of thousands – whether plucked from rooftops by helicopter, ferried in fleets of small boats, or bussed out of the disaster zone just in the nick of time. It was a haphazard rescue effort and too little too late. It took the authorities almost a week to get everyone out, restore order and get the situation under control. One year on, 1,118 people were confirmed dead and a further 135 still missing, presumed dead. Katrina left millions homeless, and some 400,000 residents left New Orleans, never to return. President George W. Bush noted the recovery would take years in a speech made on 31st August from the White House:

> The folks on the Gulf Coast are going to need the help of this country for a long time. This is going to be a difficult road. The challenges that we face on the ground are unprecedented. But there's no doubt in my mind we're going to succeed. Right now the days seem awfully dark for those affected – I understand that. But I'm confident that, with time, you can get your life back in order, new communities will flourish, the great city of New Orleans will be back on its feet, and America will be a stronger place for it.[3]

Margin of Safety

What happened could have been foreseen. A storm the size and strength of Katrina had been predicted for some time, and questions about the resilience of the flood protection systems had been raised for many years before Katrina struck. In 1998, during Hurricane Georges, waves on Lake Pontchartrain almost reached the top of the levees. John McQuaid and Mark Schleifstein of the *New Orleans Times-Picayune* wrote in 2002: 'A stronger storm on a slightly different course... could have realized emergency officials' worst-case scenario: hundreds of billions of gallons of lake water pouring over the levees into an area averaging five feet below sea level with no natural means of drainage.' This was three years before Katrina. So, what caused the levees to break? The National Science Foundation (NSF) called it 'the single most costly catastrophic failure of an engineered system in history'. With any complex system, if one part fails, the whole system collapses. Extensive investigations carried out afterwards showed that there were fundamental flaws in the design of the levee system and the way they had been built. According to a report by the American Society of Civil Engineers published in 2007, the levees were breached due to a 'combination of unfortunate choices and decisions made over many years, at almost all levels of responsibility'.[4] Some were poorly designed, making them susceptible to 'overtopping', where water pours over the top and erodes the structure away, causing it to collapse. This can't be prevented during a flood, but they should have been armoured or protected against this kind of erosion. Engineers also didn't properly consider the variability in soft soils beneath and next to the levees and how they would respond to the extreme forces involved in a catastrophic flood. The most shocking error was the safety factor considered acceptable by the levee builders. A safety factor tells you how much *extra* load a system can handle before it fails or breaks. For example, the safety factor for most passenger elevators is twelve, so if an elevator car is designed to carry up to a maximum load of ten people, or 700 kg, then the cables holding it will be engineered to be capable of carrying 120 people, or 8,400 kg (twelve times the mass of a fully loaded car). The safety factor chosen by the design engineers for the 17th Street Canal levee and floodwall in New Orleans was 1.3, extremely low by engineering standards and woefully inadequate for a critical life-saving structure. In the ten years following Katrina, federal, state and local governments spent more than $20 billion on 350 miles of new levees and flood defences designed to protect New Orleans from the kind of storm that happens once every hundred years. When Hurricane Ida passed close to the city in 2021, the defences held up, so perhaps lessons have been learned.

The failures of Katrina 'forced Americans to rethink vulnerability and risk assumptions' and prompted a cascade of reforms by the US government in how emergencies are managed.[5] Congress transformed FEMA from top to bottom, making it responsible for national preparedness, increasing its budget and giving it more autonomy. The Post-Katrina Emergency Management Reform Act (2006) also required communities to have a workable disaster plan and to ensure that local officials are specially trained to work with state and federal officials when disaster strikes. Since then, FEMA has spent more than $2 billion to train and prepare local authorities, meaning responders have better training and have put in place disaster plans in which they have confidence. National and local emergency services now go through the same disaster training and work from a single playbook, so they speak the same language and can work seamlessly together. Communities are better prepared for a worst-case scenario: emergency supplies are already in position, and hospitals and care homes are better equipped to evacuate. Another major change since Katrina is involving the public in both preparedness and response. Everyone has a role to play, not just emergency services and officials. 'That is probably the biggest change in recent years,' says William Carwile, FEMA's associate administrator for response and recovery until he retired in 2013. 'The realization that disaster response is not just a government response, it is a societal response. The federal government has a role, and so does everyone else.'

Katrina was a natural disaster made worse by a catastrophic failure of engineering. Yet it also revealed a shocking lack of preparedness, compounded by incompetence, delays and miscommunication that showed the inability of the federal, state and local governments to work together. Ten years after the disaster, President Barack Obama told a crowd in the Lower Ninth Ward, a predominately Black neighbourhood of New Orleans devastated by Katrina: 'What started out as a natural disaster became a man-made disaster – a failure of government to look out for its own citizens.' It is impossible to predict everything that might go wrong with such a catastrophic event as a Category 5 hurricane. The only sure way to keep people safe is to get them out of harm's way as quickly as possible. A successful evacuation relies on good preparation, clear communication and decisive leadership. With Katrina, Amtrak trains that could have transported people to safety well before the storm arrived were left unused, the 82nd Airborne Division spent days on standby waiting for orders that never came, and buses that were dispatched to evacuate communities went to the wrong place or never arrived at all. Every citizen in the disaster zone needs to know what the emergency plan is and what they should do if the worst happens. With Katrina, no one knew how they would get people out, there was an almost complete breakdown in communication between critical agencies involved and leaders failed to make key decisions early enough. If an efficient and well-rehearsed evacuation had been carried out on Friday, 26th August, a tragedy could have been avoided.

Chapter 6

The Great Hack

There was truth and there was untruth, and if you
clung to the truth even against the whole world,
you were not mad.

– George Orwell, *1984*

Trump's presidential election victory in 2016 was the biggest upset
in the history of US elections, a shock event that caused global
stock markets to plunge overnight. A billionaire businessman who
had never held elected office before harnessed the hopes and fears
of 'Middle America' to become the 45th president of the USA,
confounding political experts. Before the election, national polls
all had Hillary Clinton ahead by up to fifteen percentage points,
indicating an almost certain victory for the former First Lady, US
Secretary of State, and the first woman to win a US presidential
nomination. As US voters tuned in to watch the live coverage of
election night on Tuesday, 8th November 2016, the exit polls all
showed a strong Clinton lead. But as the night unfolded, it turned
out to be nail-bitingly close, with many critical states staying too
close to call until late into the night. The election was on a knife
edge. After winning Florida, Trump edged ahead in Pennsylvania,
Wisconsin and Michigan, eliminating any chance of a Democrat
win. At around 2 a.m., Clinton called Trump to concede the race.
Shortly afterwards, Trump delivered his victory speech at the
Hilton Hotel in Midtown Manhattan, telling his supporters, 'No
dream is too big, no challenge is too great. Nothing we want for
our future is beyond our reach.' The Dow Jones index dropped

850 points (further than it did after 9/11), Japan's Nikkei fell by 1,000 points and Canada's immigration service website crashed under the weight of people trying to flee the country. Everyone underestimated Donald Trump, but his rhetoric tapped into popular anxieties about the present and fury against the Washington 'swamp', echoing the golden age of the 1950s and 1980s with its vision to 'Make America Great Again'. He motivated his supporters in droves, in contrast to a Clinton campaign that seemed lacklustre by comparison. But did he also have a secret weapon that tipped the balance where it mattered most?

In America, the president isn't elected by a popular vote of its citizens. Instead, the Founding Fathers created an electoral college containing 538 'electors', each with a vote. To become president, you need to achieve at least 270 votes, or more than half. Each state is allocated a number of electoral votes depending on the size of its population, so states like California and Texas have the most votes (currently having fifty-four and forty respectively). Most states have a winner-takes-all system where the candidate that gets the most votes in that state is awarded all of their electoral votes. So, even if you win California by just one vote, you gain fifty-four Electoral College votes. So, razor-thin margins can have a significant impact on the overall result. Although Trump won the electoral college, Clinton gained more votes overall, winning the popular vote by 2.86 million, with 48.1 per cent of Americans voting for Clinton compared to 46 per cent for Trump. Trump scored narrow victories in a few vital states, making it the fifth time in US history that the popular vote winner failed to take the election. In Michigan, Trump won by just over 10,000 votes, in Wisconsin by just over 22,000, in Pennsylvania by just over 67,000. Trump won the Electoral College race with 306 electoral votes. Pennsylvania was worth 20 electoral votes, Michigan 16 and Wisconsin 10. If Clinton had won all three, she would have won the election. So, the entire presidential race came down to just 90,000 voters in just three states (out of 139 million who voted). In a victory speech afterwards, Trump described election night:

It began with phony exit polls… it looks really bad… So, it's election night and I figured we were gonna lose… then the real numbers started coming… Then it happened folks, out of nowhere… The map was bleeding red… And I'll never forget the guy who was saying for months: 'there is no path to 270 for Donald Trump'… The Electoral College is genius… it's a very different way of campaigning.[1]

Those three states were known to be battlegrounds that could swing either way long before the election. Both campaigns targeted them with aggressive advertising on billboards, TV, radio, and social media. But one campaign had a massive advantage – a massive set of data that gave deep insights into the personalities of specific voters. This allowed the campaign to design hyper-targeted social media ads that would press just the right psychological buttons to manipulate voters in those three states.

In March 2018, whistle-blower Christopher Wylie released a cache of documents to the *Guardian* revealing that political consulting firm Cambridge Analytica had gained unauthorized access to the personal private data of 87 million Facebook users and used it to design uniquely tailored advertising campaigns that might have helped the Trump campaign sway the election. A twenty-nine-year-old with bright pink hair, dark glasses and piercings, Wylie had been a data scientist at the firm for five years, and his revelations caused the biggest privacy scandal in history. Facebook's share price plummeted, government investigations on both sides of the Atlantic were triggered and Cambridge Analytica went bankrupt. Mark Zuckerberg was forced to appear before Congress to apologize and explain what had happened. Facebook was later fined $5 billion by the Federal Trade Commission, the largest penalty ever imposed on a company for violating its users' privacy. Hillary Clinton told reporters:

There was a new kind of campaign that was being run on the other side, that nobody had ever faced before. Because it wasn't just all about me. It was about how to suppress voters who were inclined to vote for me… a massive propaganda effort to prevent people from thinking straight, because they're being flooded with false information.

So, how did they acquire all this data, what was it used for and why was it so controversial?

Wylie broke it down for the *Guardian/New York Times* team working on the story. Step one was an app called 'This Is Your Digital Life', developed by Dr Aleksandr Kogan, a Cambridge University academic, in 2014. It asked US voters to complete over 120 questions to generate a detailed psychological profile of them. It's hard to convince people to fill in a survey, so a cash incentive was offered, and 32,000 people initially took it up. To get paid, the user had to log in to Facebook and approve access to their account. The app then harvested as much data as possible about the user, including personal information such as their real name, location and contact details. But it didn't stop there. It also did the same thing for all the friends of the user who had logged in – turning data from a few thousand people paid to fill in a survey into data from a few million people across the whole of the United States. The users' friends hadn't completed the personality test, but a colleague of Kogan's had discovered a way to back-engineer the personality test profiles from this Facebook data: it turns out that whether you choose to like pictures of mountains, cats, or people says a lot about your personality. This produced an enormous set of records with personality profiles attached to them. Cambridge Analytica's data scientists then created a model that could make scarily accurate predictions about the personalities, political affiliations and triggers of millions of voters. In total, 253 algorithms were developed, which meant there were 253 predictions for each record. The data was combined with other sources, such as voter records and consumer data, to create a superior set of records. According to Wylie, by August 2014, they had the first successful outputs: 2.1 million profiled records from eleven target states. The 253 predictions were the 'secret sauce' Cambridge Analytica offered its customers and were first deployed in the Republican campaigns leading up to the 2016 presidential primaries.

While Cambridge Analytica breached Facebook's rules by harvesting data without users' consent, the way the data was used represented a big leap forward for data analytics – taking available data to generate valuable insights into people's habits and behaviour as well as to exert influence. This is nothing new: political campaigns have always targeted voters, sometimes in dubious ways, and often using the latest innovation or technology to do it. The 'stump speech' was born in 1896 when William McKinley began to address the public directly, making speeches from his front porch in what became known as the 'front-porch campaign'. Other candidates preferred to travel the country by railroad, making speeches from the back of a locomotive carriage, giving birth to the phrase 'whistle-stop tour'. In 1932, Franklin D. Roosevelt visited forty-one states by train and made hundreds of stops to speak to potential voters, helping sweep him to victory. The invention of television allowed for a new type of political advertising that could reach right into people's homes, and the first televised presidential debate took place in 1960 between John F. Kennedy and Richard Nixon. Bumper stickers, catchy jingles, billboard posters, cartoons and smear campaigns have all been used to get people to vote for a particular candidate (or to encourage them not to vote for their opponent). More recently, campaigns have relied on direct mail databases, telephone appeals and websites using modern marketing techniques to better understand and target voters. For a long time, pollsters have used segmentation to help them better target likely voters by categorising them by characteristics such as age, gender, income, education, hobbies and location. The Clinton campaign in 2016 used this kind of information and had an advanced data analytics operation based on a top-secret algorithm named 'Ada' after the pioneering nineteenth-century mathematician Ada Lovelace. Psychographics takes this one step further, grouping people by values, political opinions, aspirations and, crucially, their psychological profile. This was the powerful new weapon introduced by Cambridge Analytica in the 2016 election.

Two people with the same basic demographic make-up – say white middle-aged men who live in Wisconsin, play golf and have an annual income of $80,000 a year – might think very differently. They could have different views, different personalities and very different ways of reacting to the same political message. But if you know their psychological profile – whether they are introverted, open, argumentative, analytical, conscientious, neurotic and so on – you can tailor your message perfectly to that individual, getting the response you want. With all this Facebook data, you can design adverts in such a way that everyone sees a slightly different version based on their profile. This allowed the Trump campaign to create millions of highly targeted adverts on key issues such as immigration, gay marriage, abortion and the economy, as well as messages about their opponent. Speaking to the *Guardian*, Wylie explained, 'If you're talking to a conscientious person, you talk about the opportunity to succeed and the responsibility that a job gives you. If it's an open person, you talk about the opportunity to grow as a person. Talk to a neurotic person, and you emphasize the security that it gives to their family.'

The Trump campaign's social media operation was incredibly sophisticated. Before the election, CNN reported that the campaign had paid Cambridge Analytica $5 million in September alone to tailor Trump's political ads to voters' personalities, saying this was their 'secret weapon'. In total, they ran 5.9 million ads on Facebook and spent $44 million from June to November. Hillary Clinton's campaign ran only 66,000. Shortly after the Cambridge Analytica scandal broke, Trump seemed to gloat about it, tweeting: 'Remember when they were saying, during the campaign, that Donald Trump is giving great speeches and drawing big crowds, but he is spending much less money and not using social media as well as Crooked Hillary's large and highly sophisticated staff. Well, not saying that anymore!'

Investigative reporters from *Channel 4 News* later exposed Cambridge Analytica's role in more detail. Secretly recorded interviews with CEO Alexander Nix and other senior executives showed them boasting about playing a pivotal role in bringing Donald Trump to power, using 'unattributable and untrackable' digital advertising, including the infamous 'Defeat Crooked Hillary' campaign. The meetings also included Mark Turnbull, the managing director of CA Political Global, who described how, having obtained damaging material on opponents, Cambridge Analytica could discreetly push it onto social media and the internet. He said:

> [W]e just put information into the bloodstream of the internet, and then watch it grow, give it a little push every now and again… like a remote control. It has to happen without anyone thinking, 'that's propaganda', because the moment you think 'that's propaganda', the next question is, 'who's put that out?'

Speaking to reporters posing as potential clients from Sri Lanka, Nix said he had a close working relationship with Trump and claimed Cambridge Analytica won the election: 'We did all the research, all the data, all the analytics, all the targeting. We ran all the digital campaign, the television campaign and our data informed all the strategy.' The company's head of data, Alex Tayler, added: 'When you think about the fact that Donald Trump lost the popular vote by three million votes but won the electoral college vote that's down to the data and the research. You did your rallies in the right locations, you moved more people out in those key swing states on election day. That's how he won the election.'

Wylie also claimed that Cambridge Analytica used Russian researchers and openly shared information on its campaigns with companies linked to Russian intelligence, such as Lukoil, one of their clients and Russia's second-largest oil company. Moreover, Aleksandr Kogan, who developed the data-gathering app used by the firm, also worked for St Petersburg State University. The 2019 report into Russian interference in the election conducted by US Special Counsel Robert Mueller revealed that a 'troll farm' based in St Petersburg had been active since 2014 with a mission to 'conduct information warfare' by creating thousands of fake social media accounts. These fake accounts, purporting to be Americans, spread propaganda supporting Trump and false and disparaging messages about Clinton. Throughout 2016, they posted divisive content about topics such as Black Lives Matter, immigration and gun control. They sponsored political ads criticizing Clinton and pumped out hashtags like #Hillary4Prison and #TrumpTrain to millions of followers. According to Clinton:

> The real question is how did the Russians know how to target their messages so precisely to undecided voters in Wisconsin or Michigan or Pennsylvania – that is really the nub of the question. So, if they were getting advice from say Cambridge Analytica, or someone else, about 'OK here are the twelve voters in this town in Wisconsin – that's whose Facebook pages you need to be on to send these messages' that indeed would be very disturbing.

A pink-haired whistle-blower had stunned the world with the revelation that the personal data of millions of people had been used to unleash a 'psychological warfare tool' on US voters that might have helped propel Donald Trump into the White House. But it affected the data of people from all ends of the political spectrum. It was a wake-up call to everyone that our data wasn't safe and even trusted brands like Facebook weren't protecting its users. In the immediate aftermath of the *Guardian/New York Times* stories, Facebook became embroiled in the biggest scandal in its history, and a global data privacy movement began as the hashtag #DeleteFacebook exploded throughout social media in protest at the way people's data had been stolen. Five days after the scandal broke, Facebook CEO Mark Zuckerberg apologized, calling it a 'major breach of trust' and vowing it would never happen again. While many called Zuckerberg's initial response 'too little too late', he has since put privacy at the heart of his company's future strategy. In March 2019, he published a 3,000-word manifesto on privacy via his personal Facebook account. It described a new 'privacy-focussed vision for social networking' and set out six principles for future product development at Facebook. It emphasized the importance of enabling private communication, end-to-end encryption, secure data storage, the ability to delete or 'expire' photos and video we don't want to linger on the internet for years, and the need to allow users to work across platforms regardless of who owns them. Tapping into the mood of consumers in the post-Cambridge Analytica environment, he acknowledged that 'privacy gives people the freedom to be themselves and connect more naturally, which is why we build social networks'.

The Facebook-Cambridge Analytica scandal showed that most people are entirely unaware of how precious their personal information is or how vulnerable it is to misuse. Our data is part of our digital persona, a precious asset in the online world that needs to be kept as safe as the wallet or purse in our bag. While the news media reported widespread privacy concerns and protests encouraging people to take action, very few people actually deleted their Facebook accounts or even changed their privacy settings. In the months following the scandal, researchers from the University of Bath conducted interviews with UK students to ascertain their understanding of online privacy, how they manage it, and how recent events had influenced them. They found that the people they interviewed were confused about how social networks like Facebook use their data and lacked knowledge about how it could be misused. They just didn't understand or think much about their online privacy. Not surprisingly, the interviewees expressed little concern about the issue and felt they were immune to targeted advertising, saying 'it wouldn't happen to me'.[2] Voters in the United States who felt manipulated during the election might have reacted differently, but this study highlights the fact that many of the risks relating to our behaviour on social media happen beneath the surface, they are hidden from us, and naturally we tend to ignore them. As we live our lives more and more in a digital universe, new risks are emerging, and the biggest is to our personal data. If we are to keep ourselves safe in the digital world, we need to know much more about the data we give away, what happens to it, and how it could be used to harm us.

The New Oil

When George Orwell wrote 'Big Brother Is Watching You', he was describing a dystopian future in which the state harnessed the latest technology – cameras, television and radio – to monitor and control the public. Seventy years after he wrote *1984*, the world's citizens now live most of their lives in a digital world dominated by a few social media empires that use algorithms and vast armies of human moderators to keep a watch on their users 24/7. They are able to see everything we do or say and can censor users or erase them entirely if they are found guilty of 'thoughtcrimes'. The surveillance society we now live in goes way beyond anything Orwell could have imagined. The number of people connected to the internet grew by 97 million in the twelve months to January 2024, taking the total to 5.35 billion worldwide, roughly 66 per cent of the global population. Whether on a laptop computer or mobile phone, our lives are increasingly spent immersed in a digital world: the average internet user spends six hours and forty minutes online every day. We carry smartphones with us everywhere we go, constantly shedding tiny fragments of data about our location and behaviour that are gobbled up by private companies so they can build incredibly detailed profiles of individual users. And it's not just smartphones – every connected device we use captures data in vast quantities about who we are, what we did and what we like. Every website we visit, app we download, TV channel we surf or email we send generates data that can be captured, stored and sold.

Our data is precious and has real value because it is useful. In 2006, data scientist and mathematician Clive Humby said:

Data is the new oil. Like oil, data is valuable, but if unrefined it cannot really be used. It has to be changed into gas, plastic, chemicals, etc. to create a valuable entity that drives profitable activity. So must data be broken down, analysed for it to have value.

Like oil, it powers a new industry and yet many people have no idea how it is extracted or what happens to it afterwards. A huge ecosystem of advertising technology companies extract, gather, sift and organize personal data to give powerful insights into what people will buy, who they vote for, how much money they have, where they live, what health problems they have and much more. These businesses offer a service that makes sense of vast screeds of data and builds scarily accurate psychological profiles of millions of people that can be used to target advertising in a way that is so precise and manipulative that many people are convinced they are being listened to through their mobile phone. Not only can they convince people to buy particular products, but they can also be used to get people to vote a certain way or even to change their beliefs altogether. The volume of data that exists in private databases, on the internet or on the Dark Web is now so overwhelming it is almost impossible for the human brain to conceive. It has become a precious commodity that is fuelling a booming adtech industry that is expected to be worth $2.9 trillion by 2031. The casualty is privacy. Once our personal information is out there, it is impossible to get it back. It gets collected, bought, sold, copied, exploited, leaked, stolen, hacked and archived. Some of it is given away willingly when we accept cookies or agree to share our information, some of it is accidental (who reads the terms and conditions on a new app you've downloaded?), and some of it is stolen or acquired from a data breach we might not even be aware of.

Early in 2024, cybersecurity experts discovered the biggest data leak ever, in which twenty-six billion records from sites including Twitter (now X), LinkedIn and Tencent exposed personal information in what has been dubbed the 'mother of all breaches'. Major data breaches like this happen almost once a year, and they are getting bigger, giving criminals access to people's personal information and putting millions of people at risk of identity theft and sophisticated phishing attacks. Most companies have fallen victim to some kind of data breach, with IBM reporting that 83 per cent of organizations had more than one breach in 2022. From 2018 to 2023, there was a 175 per cent increase in the number of electronic records compromised each year. Every breach is different, but a huge range of sensitive data can be exposed including names, addresses, passwords, bank account information, confidential medical records, logs of people's online activity and consumer preferences.

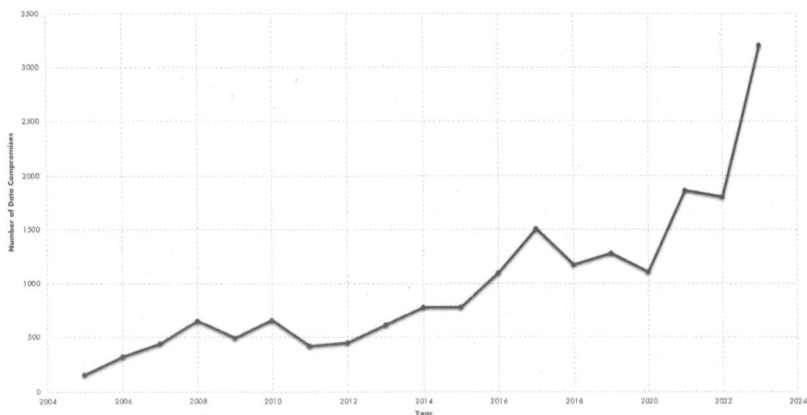

Rise in breach of personal data records in the US between 2005 and 2023
(US Identity Theft Resource Center, 2023)

As the digital world expands and AI takes root in our daily lives, hackers are becoming more inventive and cybercrime is skyrocketing. It is a threat faced equally by both businesses and individuals. Research carried out by Professor Michel Cukier from the University of Maryland found that hackers are trying to gain access to our computers all the time, with the average computer being attacked once every 39 seconds (or 2,244 times a day). They found that most attacks come from unsophisticated hackers using automated systems to attempt to guess usernames and passwords.

According to Cybersecurity Ventures, the cost of cybercrime is set to rise by 15 per cent every year, reaching $20.1 trillion by 2030. Phishing scams are the most common form of cybercrime, with an estimated 3.4 billion spam emails being sent every day to try to trick people into revealing personal information or accidentally download malware. Personal data from major breaches can be used to generate highly targeted phishing attacks that use your personal information for a scam that is so perfectly tailored that it appears genuine. The goal is usually to convince the recipient to transfer funds to a particular bank account or reveal sensitive information that allows them to steal your identity.

On the Dark Web, our data has real value to the hackers who sell it. You can buy someone's credit card number and PIN for just $5 or a social security number for just $1. Worryingly, the most valuable data being traded is a patient's private medical records, which can be worth up to $1,000. It is no surprise then that the healthcare industry is the most plagued by cyber-attacks, with one quarter of all data breaches from the health sector. Between 2005 and 2019, around 250 million people worldwide were affected by healthcare data breaches, and it is getting worse every year. Our personal medical information is probably the most private data we own and often something we do not want to be made public. But why are our medical records so valuable and how do criminals benefit from them? It is because they contain a rich source of personal information, including names and addresses, social security numbers, bank accounts, medical insurance details, demographics and much more. Unlike credit cards, which can be cancelled, medical records have a long lifespan, and criminals can use them for extended periods without being detected. As the ultimate source of personal data, medical records provide the best raw material for identity theft. Hackers can also impersonate the victim to receive medical treatment and prescription drugs or commit insurance fraud. Incredibly sensitive information on sexual activity, mental health or addiction could be used to blackmail victims. Healthcare companies are often held to ransom if they want to get the data back (or have it deleted).

In October 2022, one of Australia's largest medical insurers, Medibank, fell prey to a cyberattack, and Russian hackers obtained 200 Gb worth of data on 9.7 million Medibank customers. Australia's Home Affairs Minister Clare O'Neil described it as 'the single most devastating cyber-attack we have experienced as a nation'. They demanded a ransom of $9.7 million (USD) or they would leak the data on the Dark Web. Medibank didn't have any cyberattack insurance and so decided to follow the Australian government's recommendations and refused to pay. What happened next shocked Australians and provoked the government to direct its outrage towards Russian officials. The hackers released a set of files, named the 'good-list' and the 'naughty-list', containing highly sensitive health claim details including treatment for mental health problems or addiction and abortion records. 'Literally millions of people having personal data about themselves, their family members, taken from them and cruelly placed online for others to see. These people are cowards and scumbags' said O'Neil. In January 2024, thirty-three-year-old Russian national Aleksandr Ermakov was named and sanctioned by Australia for his role in the attack. The first time new cyber sanctions had been used, they imposed a financial sanction and a travel ban on Ermakov, believed to be a member of the Russian ransomware gang 'REvil'. The sanction makes it a criminal offence, punishable by up to ten years' imprisonment and heavy fines, to provide assets to him. While this might help satisfy a few embattled officials, the data was never returned or deleted and is still archived on the Dark Web for anyone to access, leaving victims exposed and helpless.

Online fraud is one of the world's most costly illegal activities, making criminals around $600 billion per year (almost 1 per cent of the world's total GDP), most of which comes from scams eliciting people's bank account passwords or credit card information. Worldwide, 45 per cent of internet users say they worry about scams and other types of fraud online. It is a particularly acute concern for the elderly, who are less technologically literate and tend to be most vulnerable, with 55 per cent of the over-65s worried about it.[3] Singapore is an interesting example, having both an ageing population and a high degree of personal wealth to protect, given it has the highest average income in Asia.[4] Residents aged sixty-five and above make up almost one-fifth of Singapore's population, up 11.7 per cent in the last decade, according to the Singapore government's 2023 population report. Financial concerns topped the 2023 Singapore Risk Barometer carried out by the Institute for the Public Understanding of Risk at the National University of Singapore (NUS). This showed that inflation was the most dominant risk for Singaporeans, while 80 per cent of people were also worried about increases in prices for goods and services, and 69 per cent were most worried about the risk to their investments and losing money through scams. Perhaps reflecting an older and wealthier population, these financial risks were all much higher than concerns over other prominent issues such as health (cancer, heart disease, diabetes) and climate change.[5]

People everywhere are becoming increasingly worried about personal data being stolen and the harm this could cause both to individuals and wider society. In 2021, the Lloyd's Register Foundation *World Risk Poll*, conducted by Gallup, investigated public concerns about the misuse of personal data. They asked a cross-section of the public in 121 countries about the negative consequences of the technology they use daily. They found that two-thirds (77 per cent) of internet users were worried or very worried that their personal information online would be stolen and used by companies for marketing purposes or by their government without permission. These fears were higher among people with lower incomes, and most people around the world also worry that their personal data could be used to discriminate against them. Indeed, fifty-four per cent of people who had experienced multiple forms of discrimination in the past – whether due to their gender, ethnic group or skin colour – said they were 'very worried' about their personal data being stolen online (compared to 42 per cent among those who had not experienced some form of discrimination before). 'The online world always mirrors and frequently amplifies offline harms and inequalities,' says Carlos Iglesias, an expert on web standards and open data with the World Wide Web Foundation. 'Marginalized groups are more acutely affected by safety and privacy issues. The *World Risk Poll* report... demonstrates how privacy challenges contribute to deepening inequality and make the web feel like a less welcoming place.'

The Privacy Paradox

Not only do we need to protect our data from unauthorized use, but we also need to ensure we keep our data private and are able to control who can access it and what they are able to do with it. In the Digital Age, when much of our lives are played out on the internet or on social media, privacy has never been more important. History teaches us that privacy violations often cause the most harm to marginalized groups such as Black people, women, immigrants, religious minorities, members of the gay community and others. For them, the right to privacy is a matter of survival, a way to keep themselves safe from harm and prevent discrimination. In the 1950s and 1960s, the US government gathered data and intelligence on Black Americans and used it to track people fighting racism. During the HIV/AIDS epidemic, sufferers were fearful their employers would find out from their healthcare provider and be bullied at work or even lose their jobs. More recently, since the overturning of *Roe v Wade*, which protected a woman's right to have an abortion in the US, judges have convicted abortion-seekers on evidence using people's location data, text messages and online activity.

We know that our data is used to feed powerful algorithms that deliver personalized adverts to our social media feeds or engineer what we see and experience online. Sometimes it is useful, recommending a book we might enjoy or a bargain holiday we just happen to be in desperate need of. But algorithms that profile users and target content to them can also facilitate discrimination based on age, gender, race and other characteristics. An investigation by the independent non-profit ProPublica showed that Uber was using Facebook to advertise exclusively to young men, excluding women and older people.[6] They also found a Michigan-based trucking company that specifically targeted men interested in college football and a community health centre looking for nurses that limited its audience to women only. Similar examples can be found in the housing sector, where housing applicants have been excluded based on their race, gender and sexual orientation. It has also been discovered that online mortgage lenders charge African American and Latino borrowers much higher interest rates than White borrowers with a similar credit score. A University of California, Berkeley study carried out in 2018 found that online lenders earn 11 to 17 per cent higher profits on loans to African Americans and Latinos, charging those homebuyers half a billion dollars more in interest every year.[7]

Data privacy was thrust into the spotlight by the Facebook-Cambridge Analytica scandal but remains surrounded by confusion and misinformation. Popular discourse focuses on the most newsworthy issues and trends, such as targeted advertising, government surveillance and cybersecurity, which are all important, yet the full spectrum of potential harm is poorly understood. In 2021, experts tried to classify all the ways data could be used against people. The list was eye-opening and included bullying, ridicule, voyeurism, aiding physical violence, sanctions, blackmail, political oppression, domestic abuse, workplace surveillance, discrimination, silencing political opponents, discrediting people, torture, fraud, sexual predation and sex crimes, political persecution, identity theft, predictive policing and organized crime.[8] Surveys consistently show that people value online privacy very highly and are extremely concerned about digital safety. An Ipsos poll conducted in 2022 found that an 'overwhelming majority' (84 per cent) of Americans are concerned about the safety and privacy of their personal data on the internet.[9] A similar survey by Pew Research found that 79 per cent of Americans are very or somewhat concerned about the way companies or the government is using their data.[10] This is a worrying trend, given that the right to privacy is enshrined in the US Constitution and considered an intrinsic American value. Supreme Court justice Louis Brandeis proclaimed in an 1890 *Harvard Law Review* article that Americans enjoyed a 'right to privacy', which he argued was the 'right to be let alone'. Although the word 'privacy' doesn't appear in the Constitution, it can be inferred from various sections of the Bill of Rights – the first ten amendments of the US Constitution, which set out Americans' rights in relation to their government. The Fourth Amendment contains the idea that citizens are 'to be secure in their persons, houses, papers and effects, against unreasonable searches and seizures'. Moreover, the First Amendment protects freedom of religion, speech, press and assembly. A champion of privacy rights, Brandeis famously wrote a passionate and eloquent dissent in a case on the wiretapping of private telephone conversations by federal agents:

> The makers of our Constitution undertook to secure conditions favorable to the pursuit of happiness. They recognized the significance of man's spiritual nature, of his feelings and of his intellect. They knew that only part of the pain, pleasure and satisfactions of life are to be found in material things. They sought to protect Americans in their beliefs, their thoughts, their emotions and their sensations. They conferred against the government, the right to be let alone—the most comprehensive of rights and the right most valued by civilized men.[11]

While people claim they care about their digital privacy, their behaviour tells a different story. They are willing to give away their personal information for relatively small rewards or to attract attention on social media. In a paper entitled 'Your Browsing Behaviour for a Big Mac', Juan Pablo Carrascal and colleagues from the University of Barcelona reveal that internet users value their browsing history at about seven euros, the equivalent of a

Big Mac meal.[12] The inconsistency of attitudes towards privacy and people's actual behaviour in the digital realm is often called the 'privacy paradox'. Despite worrying about digital safety, many Americans engage in behaviours that could put their online information at risk, such as using the same password for many different accounts, noting passwords in their phone, sharing passwords or only changing them when they have to. Two in three Americans (65 per cent) say they reuse passwords for different online accounts, while one in five use passwords considered familiar or easy to guess, incorporating names and birthdays.[13]

In the US, internet users are concerned and confused and feel they lack control over their personal information. According to Pew Research, 81 per cent of Americans feel they have little or no control over data collected about them by companies and the government, while 78 per cent say they have no understanding of what the government does with their data. Ironically, as users are given more choice and companies develop better ways to allow people to manage their personal data, researchers have found that 'privacy fatigue' is becoming more prevalent. Constant demands to update your privacy settings and increasingly complex privacy options and policies are causing people to feel a loss of control. Frequent news headlines about data breaches contribute to a sense of futility, making users weary of thinking about online privacy or making decisions, resorting to automatic behaviour such as clicking 'accept all' without thinking about the implications.[14] If we want to keep ourselves safe and maintain our privacy, we need to fight against this kind of apathy and support our deep-rooted belief in the right to privacy with real action. In our digital lives, we must walk a careful line that balances constant vigilance and excessive paranoia.

Technology has moved on since the 2016 presidential election and one of the biggest risks to democracy now comes from generative AI (artificial intelligence) and other tools that can create and spread disinformation much more alarming than targeted advertising. Digitally manipulated content, colloquially known as deepfakes, is a terrifying new development and, if perfectly timed, has the potential to dramatically influence elections. An AI-generated video of Donald Trump offering climate change advice to the people of Belgium was actually created by the political party Socialistische Partij Anders, but many believed it was real, and it caused a social media storm amongst Belgian voters. An early example, deliberately engineered by online news source *BuzzFeed*, showed Barack Obama on camera appearing to say, 'President Trump is a total and complete dipshit.' Another fabricated video showed Ukrainian President Volodymyr Zelensky surrendering to Russian President Vladimir Putin but was quickly detected and removed from Facebook and YouTube. While these early examples were easy to spot as fakes, the technology is developing rapidly and has already surpassed all expectations (or fears). In a new and rapidly growing scam, hackers use AI to clone your voice and fake panicked calls to friends and family asking for money. Some people have even had fake kidnapping calls where their 'child' has called them, sounding in distress and saying they need millions of dollars or they won't release them.

The same technology could have devastating consequences for democracy around the world and destabilize fragile or divided communities. The World Economic Forum *Global Risks Report* for 2024 highlights misinformation or disinformation as the most severe short-term global risk due to its potential to 'radically disrupt electoral processes in several economies'. In the two years between 2024 and 2026, around three billion people were expected to vote in national elections not just in the United States and UK but also in Indonesia, Mexico, India and Bangladesh. Generative AI with interfaces that anyone can use, even without technical skills, has caused an explosion in falsified information and 'synthetic content' such as deepfakes, sophisticated voice cloning and counterfeit websites. This kind of fake content thrives on social media and is indistinguishable from human-generated content, making it hard to identify, track and control. It can easily be used to manipulate voters and contributes to 'truth decay' – a growing distrust of information that deepens polarized views and amplifies societal division. The WEF report warns it could 'destabilize the real and perceived legitimacy of newly elected governments, risking political unrest, violence or terrorism and a longer-term erosion of democratic processes'. Unless governments work out how to regulate these new technologies or we develop ways to guarantee the source and quality of the information we see, we will increasingly live in a world where it is impossible to distinguish the genuine from the fake or facts from propaganda. Our digital lives will become more and more like the fictional country of Oceania in Orwell's *1984*, where 'Newspeak' and 'Doublethink' were used to influence and control the population.

How can you avoid being bombarded by fake news and keep yourself digitally safe? Can you find out who has your personal data and get it back or erase your digital footprint altogether? A first step could be to clean up your digital history, remove old emails or files from the cloud, and get rid of social media accounts you don't use that might still contain embarrassing posts you'd rather not have in the public domain. Getting your data back or having it deleted is a bit harder. There are around 540 data brokers in the US alone, holding a vast amount of personal data about consumers. Since the Facebook/Cambridge Analytica scandal, digital privacy has become a global movement, with activists and campaigning organizations gaining concessions from technology giants like Apple while pushing governments for more protections. The Electronic Frontier Foundation (EFF) is one of the biggest and dates to the early days of the internet, being founded in 1990 to champion user privacy and free expression. Aaron Schwartz, a senior staff attorney for the EFF, says: 'It seems like every week there's a breach. To state the obvious, if the information isn't collected in the first place or stored, this wouldn't be an issue.' Asking to opt out or to delete the information they hold on you can take forever, and so some tech start-ups are capitalising on users' growing privacy concerns. A whole ecosystem of apps and services now claims to help you 'take back control of your data'. PermissionSlip is an app designed by Consumer Reports that offers to find out what information these companies hold about you and file a request to stop them from selling your data and erase your records. Other services like DeleteMe offer to find and remove your data from search engines and data-broker sites. They claim to expose an average of 2,400 data records during a two-year subscription. Rob Shavell, DeleteMe's co-founder and CEO, believes that for some risks, like your name appearing with data brokers, you 'might need to take an aggressive approach and get professional help'.

Since the Facebook/Cambridge Analytica scandal, the battle for digital privacy has already changed the internet: tough new regulations have been introduced by governments worldwide, and the way technology companies think about their customers has fundamentally shifted. Governments were outraged, and multiple investigations took place in the US and the UK, where Cambridge Analytica was involved in the Brexit campaign. In the following years, regulators toughened data protection laws and demanded change, although many doubted they had the will or the power to limit the mass harvesting of personal data. In Europe, the General Data Protection Regulation, or GDPR, came into force a few months later, on 25th May 2018, giving the authorities the power to impose fines of up to €20 million or 4 per cent of a company's revenue. California passed a law shortly afterwards that gave its residents the right to see what personal data a company had collected about them and request that it be permanently deleted. Whether due to government pressure or a shift in the mood of its customers, the industry has responded, making privacy one of its selling points. Online privacy is now the mantra for almost all of the tech giants, such as Google and Apple, which have been rolling out tools that help their users block their data from marketeers and give people much more choice and transparency every time their data is exposed. Shortly after the scandal, the Mozilla Firefox browser offered an extension that could prevent Facebook from tracking your activity, and in 2021, Apple introduced a pop-up window that asks people for their permission to be tracked by different apps.

Computer experts and technology gurus advocate better cyber-security and advise people to install antivirus software, use unique passwords for every account they have and create them with a random password generator if they can, use multi-factor authentication to give an extra layer of protection if their password is leaked in a data breach, and pay particular attention to their privacy settings on social media accounts. But, no matter how secure you make yourself, criminals adopt new technologies faster than anyone else and are constantly innovating new ways to attack unsuspecting consumers. In this arms race, the only way to protect ourselves is to become digitally literate and keep ahead of the scammers. In Thailand, grandparents are going back to school to learn how to protect themselves from online scams. Thailand has an ageing population of thirteen million people, or about 20 per cent of the Thai population, over sixty years old, which will grow to 30 per cent over the next twenty years. Research by the Intelligence Centre for Elderly Media Literacy (ICEML) at Mahidol University, with the support of the Thai Health Promotion Foundation (ThaiHealth), found that elderly people are much more likely to fall victim to scammers due to low levels of digital literacy, with 70 per cent already having been deceived into sharing their personal information. In response, ThaiHealth has set up 2,456 schools for the elderly nationwide to teach digital media skills and how to use devices or online services safely.

Education, especially for the most vulnerable groups, is critical, and people of all ages and in all communities need to learn new skills faster than hackers and fraudsters can develop new tactics. *New York Times* columnist Thomas Friedman believes, 'The future of humanity will depend on our ability to adapt to and work with technology and the way through to this goal is lifelong learning.'[13] Digital safety is something everyone will need to work at every day, resisting apathy or fatigue and being constantly vigilant. If people genuinely care about their privacy and want to protect themselves online, they need to do more than just keep up to date with the latest systems, privacy options and security issues. It requires a complete change in mindset in how we interact with the digital world, ensuring our day-to-day behaviours match our concerns and keep us safe.

Chapter 7

Food and Fear

Having enough food to survive was a priority for the earliest humans, who spent most of their time hunting or foraging for food. While it was a necessity, it was also risky, and every bite they ate exposed them to potentially deadly contaminants they couldn't see and didn't understand. Despite millions of years of progress, foodborne disease is still one of the biggest risks on the planet. One in ten people fall ill every year from eating contaminated food and 420,000 people die as a result. Children under five years of age account for one-third of those deaths, and it is most severe in the poorest countries, where food is often prepared using unsafe water, food production and storage are inadequate, there are low levels of education or awareness about basic hygiene, and food safety legislation is ignored or non-existent.[1] The risk is growing fast. Most estimates suggest we will need to feed an additional two billion people by 2050 as the world's population is set to swell to over nine billion. So, by the time you finish reading this chapter, there will be an extra 2,000 mouths to feed. The scale of the looming food crisis is immense: as more and more people adopt meaty Western-style diets, we will need to produce as much food in the next fifty years as we did in the last 10,000 years. And, as the UN's Food and Agriculture Organization has stated, 'if food isn't safe, it isn't food'.

But what do we really know about the food we eat? Where does it come from, how is it processed, and just how safe is it? While we might not know much about our food, it plays a significant role in our lives. Our relationship with food is complex and buried deep in the collective consciousness. We spend a substantial proportion of our lives thinking about food, preparing food or eating it. The average Briton spends 1 hour 7 minutes each day eating and drinking in contrast to the French who spend a leisurely 2 hours 13 minutes.[2] Food is necessary for survival, but it has become much more than mere fuel and serves a function in society far beyond its basic utility. Our relationship with it is intimate, both through the visceral act of putting food inside our bodies and the fact that it contributes so much to feelings of belonging, memory and culture. It is centre stage during some of the most significant moments in our lives: the wedding feast, anniversary dinner or funeral banquet. Dietary laws are an important part of many religions, while traditional cuisine helps define entire nations, both of which contribute to a sense of identity. Food can also create powerful memories that define a particular time and place. In À La Recherche du Temps Perdu (In Search of Lost Time), Marcel Proust is famously transported back to his childhood by the taste of a madeleine cake dipped in tea. Indeed, a 'madeleine de Proust' is now a common French expression referring to a smell, taste or sound that dredges up a long-lost memory.

Food also reassures and makes us feel like we are at home, even if thousands of miles away. In Italy, the dinner table is the symbolic centre of family life; the ritual of eating together allows Italians to maintain a close bond, reinforce their values and transmit culture to new generations. The very phrase 'comfort food' evokes the safety and familiarity of home. It is also at the core of our social interactions, whether a business lunch, dinner date or night out with friends. Epicurus believed that a simple meal shared with good friends was one of the keys to leading a happy, tranquil life characterized by *ataraxia*, meaning peace and freedom from fear. Given its importance, how do people feel about the food they eat and the potential risks it brings? Food makes us feel good, but it can also provoke extraordinary levels of panic and fear – perhaps because danger to our food strikes at the very foundations of our social fabric. As Jonathan Safran Foer writes, 'Food is not rational. Food is culture, habit, craving and identity.'[3] In the summer of 2011, Germany experienced the deadliest outbreak of foodborne disease ever recorded in Europe. Caused by a rare strain of the *E. coli* bacteria, around 2,000 people were infected in the first three months of the outbreak and eighteen people died. While the statistical risk was relatively low, the whole country became concerned about the food they were eating, and this quickly turned into fear. People expecting safe and healthy food felt threatened, and this changed the eating habits of most of the population.

The public reaction to outbreaks of foodborne disease sometimes seems irrational, especially to scientists. However, our response to risk is rarely based on numerical information alone: instincts and emotions have much more influence than facts. The fact that food is essential for survival, combined with our intimate relationship with it, means that foodborne disease strikes at the very heart of our modern lives, causing symptoms that are sometimes horrible or debilitating and undermining the safety of our home life. The risk is even scarier because the cause is microscopic, hidden from view and impossible to imagine, making it mysterious and unknown, just like a shark lurking beneath the waves. Foodborne disease is nothing new and was one of the most common causes of death among early hominids, whose diet played an important role in the evolution of the species. What different civilizations in the past experienced and how they thought about the safety of their food can give us a deeper understanding of the way people respond to food risks today.

The Caveman Diet

The caveman or paleo diet was one of the most widespread fads of the last decade, advocating a return to the eating habits of our hunter-gatherer ancestors to lose weight and become healthier. This highly nutritious, carb-free diet was dominated by a huge diversity of fruits and plants with some meat if you were lucky. But wasn't until relatively recently (around 10,000 years ago) that humans began cultivating plants such as rice and wheat, living in larger settlements and farming instead of foraging and hunting. The staple diet changed during this first Agricultural Revolution, and more calorific grains became the norm. However, the diet of our prehistoric ancestors was dictated by necessity not by choice: food was scarce, you had to eat what you could find, and our forebears typically died of poor health aged between twenty and thirty years old. They were opportunistic hunters who would happily eat meat from abandoned carcasses, and it was often the sickest, weakest animals that were easiest to catch. Because of their lifestyle, 'we can assume they regularly ate contaminated meat and fish, poisonous mushrooms, toxic plants, and raw indigestible grains'.[4] They had no way of preserving food and, in tough times, would probably have eaten decomposing meat stored after a large kill. It is reasonable to assume they probably suffered from a wide range of foodborne illnesses, some of which were no doubt fatal.

The Wonderwerk Cave, an ancient cavity in the dolomite hills that lie at the very centre of the Northern Cape province in South Africa, has been excavated and studied by archaeologists since the 1940s, revealing a rich history of early human habitation. Among their finds are the remains of communities of *Homo erectus*, or 'upright man', the first human ancestor who appeared around two million years ago and spread throughout Eurasia. Wonderwerk means 'miracle' in Afrikaans, and the cave is famous for being the place where fire was first harnessed by humans, containing the remains of prehistoric campfires that represent the earliest known use of controlled fire.[5] Archaeologists recently found charred animal bones and plant remains in a layer of rock containing hand axes and other tools, evidence that not only had they tamed fire, but they might also have used it to cook their food. Exactly when humans first began cooking their food and the impact this had on the evolution of the species is a topic of enormous interest amongst anthropologists. More definitive evidence was published in 2022, showing that *Homo erectus* first cooked food around 780,000 years ago. Researchers working at a site in Gesher Benot Ya'aqov, in the northern Jordan Valley in Israel, found well-preserved ancient fish teeth near the settlement's fireplaces. By analysing the crystal structure of the teeth, the team found that they had been cooked at around 500 degrees Celsius, indicating a controlled temperature and not just burning. The authors believe this is the first concrete proof that *Homo erectus* had the ability to control fire and use it to cook food.[6] Cooking over fire eliminates many of the risks from microbes, parasites, and viruses, but not all, so our earliest ancestors would still have needed strong constitutions to survive. But cooking food might have had other benefits and could have played a critical role in accelerating our evolution. Primatologist Richard Wrangham believes that it was cooking that allowed our ancestors to transform from apelike beings to *Homo erectus*; the habit of eating cooked food rather than raw food fuelling the growth in the size of the human brain and turning us into the 'intelligent, social, and sexual' species we are today.[7]

While the advent of farming and irrigation helped feed larger populations and allowed the creation of great cities, it also allowed communicable diseases to flourish. If there was a good harvest, surplus crops could be stored to help survive the winter or drought, so early settlements began to store food but with little experience or knowledge of how to do it safely. Poorly stored grains, for example, were susceptible to contamination or mould, and illnesses like ergotism quickly became common. Ergotism, or ergot poisoning, was caused by eating grains such as wheat, barley or rye infected by a fungus, and the more you ate (perhaps over a long period), the worse the symptoms were. It was widespread in communities that relied on these grains as a staple food, in breads, soups, stews or beer and was recorded in ancient Egypt and Mesopotamia, being first described in an Assyrian tablet as a 'noxious pustule in the ear of grain'. One version of the disease, known as St Anthony's Fire, made its victims feel like they were being consumed by flames. It caused the limbs, as well as fingers and toes, to become severely swollen, and the victim to experience sensations of extreme heat (the 'holy fire') until, after a few weeks, gangrene set in, and eventually, the limbs died and fell off. Another form of ergotism causes convulsions, and symptoms that range from painful seizures, spasms, diarrhoea and vomiting to headaches, as well as mania, psychosis and delusions. It is thought that an outbreak of ergotism could explain the strange behaviour seen in the colony of Salem, Massachusetts, in 1692, which occurred after a harsh winter and damp spring, perfect conditions for ergotism in the local supply of rye. Salem physician William Griggs expertly diagnosed this as a case of 'bewitchment', leading to the execution of nineteen people found to be guilty of witchcraft.[8]

A Mysterious Death

Despite conquering much of the Western world, Alexander the Great died shortly before his thirty-third birthday, just thirteen years after becoming King of Macedonia and long before he could have enjoyed the fruits of his labour: a united Greece that ushered in the Hellenistic age, during which Greek cultural influence peaked throughout the Mediterranean and beyond. Like many of the great leaders in history, his life and death are a mystery and obfuscated by myths. His unusual and sudden death at a young age has spawned multiple theories based on the limited evidence available. Alexander died on 11th June 323 BC at the grand palace of Nebuchadnezzar II in Babylon, shortly after re-entering the city. After a series of parties and feasting, during which it is said he consumed vast quantities of wine, he collapsed, complaining of a severe pain in his back 'as though smitten with a spear'. He was taken to bed, where he developed a fever. Over the next ten days, he grew weaker, eventually losing his voice and 'lying speechless as the men filed by, he yet struggled to raise his head, and in his eyes there was a look of recognition for each individual as he passed'.[9] Shortly afterwards, he was declared dead, yet his body stayed remarkably 'fresh' for six more days before it began to decompose, adding to the legend that he was a god. Almost immediately, theories about the cause of death began to form and historians have debated it ever since. Did he drink himself to death? Perhaps he fell victim to an infection like West Nile virus or pernicious malaria. Was he assassinated or poisoned by one of his commanders? Did he die from a broken heart? But, according to new research, he might have been the first recorded death from a foodborne disease.

The Death of Alexander the Great after
the painting by Karl von Piloty (1886)

One of the most powerful and influential leaders in ancient history, Alexander was a military genius and 'remains the touchstone by which those who embrace the profession of arms measure all things'.[10] After defeating the mighty Persian empire, he established the largest kingdom the world has ever seen, stretching from Macedonia and Greece in the west to Egypt and the Indian subcontinent. In 326 BC, eight years after embarking on the conquest of Persia, Alexander was involved in his final campaign, securing the easternmost borders of his new empire in India. He was victorious over the Indian King Porus in a fierce battle involving war elephants on the banks of the Hydaspes River in Punjab but lost around 1,000 men and many were injured. Alexander's legendary horse Bucephalus, which he famously tamed aged nine years old and had ridden ever since, was killed in a cavalry manoeuvre. In the aftermath, exhausted and longing to return home to their wives and children, his army refused to go further east, so Alexander was forced to return to Persia. He split his troops, sending half to Carmania (southern Iran) and leading the rest back to Persia through the Gedrosian Desert, a long and perilous journey that claimed the lives of thousands of his men. Eventually, he made it back to his headquarters in Babylon, but not before losing his closest companion, Hephaestion, who came down with a fever and died after breakfasting on boiled fowl and a cooler of wine (against the advice of his physician). Alexander was devastated; according to Plutarch, his 'grief at this loss knew no bounds'[11] and the unfortunate doctor, Glaucias, was hanged for not taking better care of his patient.

Still in mourning, he returned to Babylon, ignoring the Chaldaean soothsayers he met on the way back who begged him to suspend his march into the city, saying that the god Belus had made an oracular declaration to them and that 'his entrance into Babylon at that time would not be for his good'.[12] When he arrived, he was met by delegations from Greece who came to 'crown and eulogize' him for his victories and to rejoice on his safe return from India. Contemporary historians describe how Alexander was 'insatiably ambitious of acquiring fresh territory' and spent the next few days frantically planning a new conquest of Arabia, overseeing the construction of a massive fleet of warships. His nights were filled with celebratory banquets and remorseful drinking binges. What kind of food would have been served at these Babylonian feasts and how safe was it? It has recently come to light that the ancient Mesopotamians were excellent cooks and left behind the oldest known recipes in a cuneiform cookbook carved in stone. They can be found in a group of clay tablets, now in the Yale Babylonian Collection, and contain thirty-five recipes composed in Akkadian and dated to the middle of the Old Babylonian Period, no later than around 1730 BC. Most of the recipes are for soups and stews and some are for preparing and cooking birds.

Together with other archaeological evidence, they tell us that the staple diet consisted of grains like barley and wheat, fruit and vegetables. Local fruits included grapes, figs, dates, pomegranates, apples, and pears. They had at least eighteen different kinds of cheese and a remarkable 300 types of bread. Olives were eaten and pressed for oil, and pickled grasshoppers were apparently a local delicacy. Fish and meat included lamb, beef, pork and venison, and game birds were also on the menu but most likely only for special occasions, or for the royal palaces. The Yale recipes reveal a significant innovation: cooking in water. They all involve combinations of meat, vegetables and grains boiled or simmered in water, which greatly increases both their digestibility and nutritional value, an improvement on more ancient forms of cooking over open fire or using heated rocks. In fact, the Babylonians employed a considerable array of cooking methods: baking in primitive ovens, over fire or hot ashes, broiling, grilling and spit-roasting. Despite this, much of their food would still have been eaten raw. Water, which came from the Tigris and Euphrates rivers, was known to contain parasites and cause disease, so wealthy residents drank wine or beer instead. The only complete recipe in the Yale collection is for a slow-cooked and aromatic lamb stew reminiscent of the kind of food the Middle East is still famous for today. Garlic, shallots, spring onions, leeks and sesame seeds, together with coriander and spices such as nutmeg or cinnamon, gave their food a distinctive flavour but could also have been used as a way of disguising poor-quality meat or meat that was beginning to rot. Indeed, adding nuts to meat dishes, a hallmark of Persian cuisine, is known to have a masking effect, suppressing the 'off' flavour of old meat.

While the Babylonians had a sophisticated menu and a wide variety of cooking techniques, they had little knowledge of basic food hygiene. They had no way to refrigerate or store food safely and sanitation was rudimentary at best. Foodborne illnesses would have been a commonplace occurrence. The idea that Alexander's death, during a long period of feasting, was caused by something in his food, is not at all far-fetched and is probably much more likely than a rare tropical disease or political assassination. Writing in the *New England Journal of Medicine* in 1998, Dr David W. Oldach gives a differential diagnosis that treated Alexander like any other patient. In it he summarizes the fundamentals of the case: a thirty-two-year-old man presents with fever and pain. He had been well until the day after heavy alcohol consumption (twelve pints of wine) and had a sharp pain in his right upper quadrant so severe that he cried out. He later had chills, sweats and fever, which continued into the next day. After listing the rest of his symptoms and noting the patient's personal history, he highlights the telling fact that after death 'signs of putrefaction were notably absent'. After considering all the possible explanations, he diagnoses the patient with typhoid fever, which is caused by *Salmonella typhi* bacteria found in con-taminated food or drinking water and can also be spread by poor hygiene. He cites a recent study showing that around 60 per cent of typhoid fever cases include sudden severe abdominal pain, which is often localized in the upper right-hand area. According to Oldach, the sharp abdominal pain was a vital clue as it indicates the disease perforated his intestine, hastening death. It also explains Alexander's last few days when he was unable to move and could only communi-cate with his eyes. A complication with typhoid fever is 'ascending paralysis', starting in the feet and moving up the body, and can also slow down breathing, eventually making a patient look dead even if they are not.[13] If he had been a patient today, Alexander could have been treated with antibiotics, although if he was living in a country with clean drinking water and a plentiful supply of safe food, his chances of contracting it in the first place would be virtually nil. If, however, he came from a low-income country in Africa, India or South Asia, he would be at risk. Despite thousands of years of medical experience, around nine million people still get sick from typhoid fever and 110,000 people die from it every year.

A Broken System

Over 200 diseases, ranging from diarrhoea to cancer, are caused by eating food contaminated with bacteria, viruses, parasites or chemical substances such as heavy metals like arsenic. An estimated 600 million people, almost one in ten people in the world, fall ill because of eating contaminated food and 420,000 die every year. It can occur at any stage of the food supply chain, from the original producer through to the end consumer, and there are many sources of potential contamination, from pollution in water, soil or air to unsafe food storage and processing. The most common are diarrhoeal diseases, which account for more than half the cases of foodborne illness worldwide and are caused by eating raw or undercooked meat, eggs, fresh produce, and dairy products contaminated by norovirus, *Campylobacter*, *Salmonella*, or *E.coli*. Amongst the most serious globally are typhoid fever, hepatitis A, cholera, and aflatoxin (produced by mould on grain stored in damp conditions). The problem is most acute in Africa and Southeast Asia which sees the highest number of cases as well as the highest death rates. Children, especially in these regions, are most at risk, with one-third of all deaths globally from foodborne diseases being children under five years of age. This is because their immune systems are still developing so they cannot fight off infections as well as adults and they also produce less of the stomach acid that kills harmful bacteria, making it easier for them to get sick.

The biggest tragedy of foodborne disease is seen in the developing world, which accounts for 75 per cent of the global deaths due to foodborne illness (despite comprising only 41 per cent of the population). Africa is by far the worst affected, with an incidence of foodborne disease per head of population that is twenty-seven times that of Europe or North America. The tropical climate of many of these countries doesn't help, providing ideal conditions for the proliferation of pests and naturally occurring toxins. Preparing food with unsafe water is one of the biggest causes, and at least 1.2 billion people worldwide are estimated to use water that is not protected from contamination by human waste. Poor hygiene and a lack of sanitation, bad conditions in food production and storage and insufficient (or absent) food safety laws are also to blame. In many of these countries, strict food standards are only applied to products destined for export and the supermarket shelves of wealthy nations. There are usually very few controls to ensure that food entering the domestic market isn't contaminated during its production, storage, transport or processing. In addition, many people in low-income countries buy their food from informal markets or sellers, many of which will lack the equipment, facilities or knowledge needed to keep food safe. Low levels of literacy and education mean many people have very little understanding of basic hygiene or awareness of the risk of foodborne illness and what to do about it.

As a country's economy develops, the situation often gets worse before it gets better. In a rapidly developing nation, the agricultural landscape changes and more intense farming practices put in place to maximize production often result in increased levels of disease. Sometimes the most nutritious foods also carry the biggest risks. Meals that come from animal products are highly beneficial yet are much more likely to cause foodborne disease. In many low-income countries, the choice of food is not only dictated by cost and availability but also by cultural practices and beliefs. In Ethiopia, raw meat called *tere siga* is one of the most popular communal dishes, while in Uganda people eat raw eggs in the belief that they cure illness. Nomadic people in West Africa believe that raw milk does not cause illness, and in Vietnam, it is commonplace to eat raw (or undercooked) blood, meat and fish, which can lead to zoonoses – infections that jump from animals to humans.[14]

Perhaps because they are relatively rare, outbreaks of food-borne disease in affluent countries that have rigorous food safety standards usually come as a surprise and cause widespread public outrage. In 2015, the owner and CEO of the Peanut Corporation of America, Stewart Parnell, was jailed for twenty-eight years after being convicted of seventy criminal charges related to an outbreak of salmonella in 2009. It was the most severe sentence ever given in a food safety case and long-awaited justice for relatives of the victims of the biggest outbreak of foodborne disease in American history. In total, nine people died and 714 fell ill (half of them children) after eating food products containing contaminated peanuts. One woman from Minnesota died after eating contaminated peanut butter. Her son, Jeff Almer, was one of the relatives who attended the court hearing, saying, 'My mother died a painful death from salmonella, and the look of horror on her face as she died shall always haunt me... I just hope they ship [them] all to jail.' The cost of such outbreaks is huge. The United States Department of Agriculture estimates that foodborne illnesses cost the United States at least $15.6 billion annually in lost productivity and medical care. Similarly, the World Bank recently found that the impact of unsafe food costs low- and middle-income economies about US$ 110 billion and a large proportion of these costs could be avoided if they adopted preventative measures to improve how food is handled from farm to fork.

Dealing with the risk of foodborne disease is much more com-plicated in countries that don't have enough to eat or where people can't access affordable food. Why would you worry about how safe your food is when you are starving? While food is vital to our existence, it was only in 2004 that 160 countries voted at the United Nations to make food – safe food – a basic human right rather than a commodity. Yet millions of people around the world still don't have enough to eat and don't know where their next meal is coming from. The scale of the global food crisis is enormous, with 793 million people estimated to be suffering from chronic hunger, meaning they are not getting enough food to lead a normal, active life. It isn't because we don't produce enough food; in fact, the world produces enough food to feed every one of the eight billion people currently on the planet. However, increasing conflict, climate change, food waste, and the rising cost of fuel and fertilizer have combined to cause hunger to start to rise after a decade of steady decline. Seventy per cent of the world's hungry people are in areas of conflict or violence, and Russia's invasion of Ukraine caused the price of wheat, grain and fertilizer to skyrocket, disrupting the supply of these vital food-producing supplies to the rest of the world. There is plenty of food around, but the problem is accessing it or being able to afford to buy it. In his ground-breaking study of famine, Nobel laureate and economist Amartya Sen noted, 'Starvation is the characteristic of some people not having enough food to eat. It is not the characteristic of there not being enough food to eat.'[15]

In 2023, the UN declared the world's food system broken, realising the impossibility of creating a world totally free from hunger by 2030. At the opening of the UN Food Systems Summit in Rome in July 2023, UN Secretary-General António Guterres said:

> Global food systems are broken – and billions of people are paying the price. More than 780 million people are going hungry while nearly one-third of all food produced is lost or wasted. More than three billion cannot afford healthy diets. Broken food systems are not inevitable. They are the result of choices we have made.

The challenge had already been included in the seventeen Sustainable Development Goals (SDGs) adopted in 2015 by all United Nations members to create peace and prosperity for people and the planet, as well as tackle climate change and preserve forests and oceans. Goal 2, ' Zero Hunger', seeks to 'end hunger, achieve food security, improve nutrition and promote sustainable agriculture'. The idea that food must be safe is encapsulated in the language, the primary target being 'for all people to have safe, nutritious, and sufficient food all year round'. Indeed, food safety is linked to achieving many of the SDGs, not just ending hunger but also eliminating poverty, promoting good health and well-being, taking climate action, providing decent work and economic growth, reducing inequalities and protecting life on land and in the oceans. So, what role does safety play in fixing the world's food system?

Food safety matters because it underpins many solutions that could fix this broken system while helping us keep pace with the exploding demand to avoid future crises. Future food needs to be safe, secure and sustainable. According to the World Bank's Safe Food Imperative, we will only have global food security when 'the essential elements of a healthy diet are safe to eat, and when consumers recognize this'.[16] The safety of food is vital to the transformation of agriculture needed to feed a growing population, as well as the development of new technologies and the modernization of current food systems. Scientists are quickly developing and scaling up production on new sources of protein, such as lab-grown meat, insects and seaweed, that could potentially feed our future population cheaply and sustainably. Advances in DNA sequencing and genomics are paving the way towards safe genetically modified crops that thrive as climate change alters global temperatures. At the same time, a new generation of consumers are driving a food revolution, giving up meat and opting for a plant-based diet many believe will help save the planet. In 2023, researchers at the University of Oxford looked at the dietary data of 55,000 individuals and found that vegans generate 75 per cent less in greenhouse gas emissions than meat-eaters. They also discovered that a meat-free diet causes significantly less harm to land, water and biodiversity.[17] Given that the entire food system is responsible for around one-third of all planet-heating emissions, perhaps we should all consider giving up the hamburger.

Aquaculture, the farming of fish and other aquatic organisms, has been the fastest-growing food production sector worldwide for many decades, and now provides us with more fish than are caught from the wild. If done well, aquaculture can provide a highly sustainable source of food while also improving the health of our oceans. The cultivation of seaweed alone could provide the miracle needed to solve the current crisis. Not only is seaweed a nutritious food, but it can also be used as an alternative to plastic, restore the oceans, and help reverse climate change. Seaweed needs only sunlight, salt water and nutrients. It grows much faster than tropical forests so absorbs more carbon per acre than any terrestrial vegetation. While much of the world is yet to embrace seaweed as a food, seaweed farming is growing fast in Asia, which is responsible for 98 per cent of the 35 million metric tonnes of seaweed sold worldwide. A study led by the University of Queensland highlighted the huge potential for seaweed cultivation, showing it could be farmed across an area of ocean the size of Australia, providing food, feed supplements for cattle, and even alternative fuels. They predict it could make up ten per cent of human diets by 2050 and would reduce the amount of land needed for food production by an area twice the size of France. In *The Seaweed Revolution*, food futurologist Vincent Doumeizel makes it clear that the age of seaweed is fast approaching:

> Spreading seaweed cultivation throughout the world... would offer us infinite potential for innovation and vastly increase the limits of our resources... Together we could enter a new era, the result of a change as pivotal as the advent of agriculture in the Neolithic period.[18]

Two things are holding back the global expansion of the seaweed industry. Convincing meat-loving Westerners to embrace it as a food isn't an easy task, despite the huge health benefits.

All seaweeds are edible (unlike terrestrial plants, none is toxic to humans), they are low in fat and are loaded with vitamins and other nutrients, including B12, which nourishes the brain. Researchers have shown that they help fight many types of cancer and contain the rare Omega-3 fatty acids that regulate cholesterol and help reduce the risk of heart disease. The benefits are well-known in countries like Japan, which celebrates National Seaweed Day on 6th February every year. During the Covid-19 lockdowns, while people in the West were stocking up on pasta and toilet rolls, the Japanese rushed out to buy dried seaweed. As a result, the obesity rate in Japan is only 4 per cent (compared to 38 per cent in the United States) and people live much longer; the average life expectancy for Japanese men is 84.5 years.

The second barrier is how to dramatically increase production safely. Of the 12,000 species of seaweed, we currently know how to cultivate around thirty, and the rest are a mystery. The rapid expansion of any new industry can result in problems, especially when there is a lack of knowledge and little or no regulation. The risks are similar to those seen in the developing world with the growth of intensive farming: outbreaks of disease, the introduction of new pests, unforeseen impact on native species or habitats, and the control of potential contaminants in the production process. The red seaweed *Kappaphycus*, also known as elkhorn sea moss, is one of the most valuable because of its carrageenan content, a product used widely in food and pharmaceuticals. Originally from the Philippines, it is now being grown in thirty countries worldwide. But the rapid growth of this highly profitable crop has been blighted by 'ice-ice' disease – a bacterial infection causing whitening of the seaweed branches. In the Philippines alone, this disease has caused a 15 per cent loss in production between 2011 and 2013 (equating to around 268,000 tonnes), representing a financial loss of over US$310 million.

Attitudes to food are deeply embedded in the public psyche but are slowly changing as the global explosion in plant-based restaurants and cafés demonstrates, driven more by environmental concerns and shifting values than by taste. Between 2014 and 2018, the number of Americans following a vegan diet increased by 600 per cent, spurred on by celebrity endorsements, concerns over animal welfare and a growing desire to reduce their carbon footprint. Albert Einstein was ahead of the times in believing that 'Nothing will benefit human health and increase the chances for survival of life on Earth as much as the evolution to a vegetarian diet.' According to Bloomberg, the worldwide plant-based food market is surging and could be worth $162 billion in the next decade, led by industry giants like Beyond Meat and the adoption of alternative foods by major global restaurant chains. But if the nascent revolution is to keep going, understanding the psychology of consumers is vital. It will largely depend on how people think and feel about some of these new foods and what influences their choices. While health and environmental impact are strong drivers, so too are perceptions about cost, taste and style. Above all, it will rely on trust: we need to know that the food on our plate is safe to eat. Because most of the food system is hidden from the average consumer, every part of it has a role to play. As we all know, trust takes years to build, can be lost in seconds, and takes forever to repair. A well-publicized food scandal could destroy public confidence for years.

Mad Cows and Englishmen

In October 1987, the *Sunday Telegraph* reported that a 'mystery brain disease' was killing the UK's dairy cows. Affected cows had trouble walking, acted erratically, and could be unusually aggressive. Over the next two years, cases skyrocketed, soon reaching epidemic proportions. The plague, dubbed 'mad cow disease' by the media, was devastating farming communities and wrecking lives. Bovine spongiform encephalopathy, or BSE, is a neurodegenerative disease found in cattle and is believed to have been caused by feed that contained the remains of dead sheep and cows (a practice banned in 1988). At its peak in 1992, 37,280 cases of BSE were confirmed in the UK. Over the course of the crisis,

4.4 million animals were slaughtered, at a cost of around £4 billion, to eradicate the disease from British shores. Based on expert scientific advice, the British government believed the chances of humans becoming infected with BSE were vanishingly small, and so, in a misguided effort to safeguard a vital industry and protect Britain's political alliances, the government adopted the mantra 'beef is safe' and went to considerable lengths to reassure the public that there was no risk to human health.

One of the most enduring political images of the early 1990s is that of Agriculture Minister John Gummer feeding a beefburger to his four-year-old daughter Cordelia at a press event intended to convince the public that beef was safe. At this point, sales of beef were down 60 per cent and there was a new scare story in the media almost every day speculating on the possible danger to humans. So the minister decided that drastic action was needed and organized a photo opportunity at an Ipswich fair on 10th May 1990. After finishing his burger, he said, 'When you have the clear support of the scientists that deal with these matters, the clear support of the Department of Health, and the clear action of the Government, there is no need for people to be worried. I shall go on eating beef, and my children will go on eating beef because there is no need to be worried.' And it wasn't just politicians reassuring the public. Britain's chief medical officer, Sir Donald Acheson, added, 'beef can be eaten safely by everyone, both adults and children, including patients in hospital'. But the public weren't convinced, and beef sales continued to fall as fears grew. Artist Roger Hiorns describes it as the moment 'the public lost their faith in the governance of this country'. The government's paternalistic attempts to reassure the public seemed to have the opposite effect, perpetuating worries and leading to new fears there may be some risk after all. Indeed, the day after this infamous press stunt, a new story about a Siamese cat found to have died from a brain disease 'similar to BSE' hit the headlines, fuelling concerns, especially among Britain's seven million cat owners.

Whether an indictment of waning public confidence in an

embattled government or a response to hyped-up stories in the media, the public was deeply worried about catching 'mad cow disease' and boycotted beef products in droves. The government believed it was acting on the best scientific evidence available at the time. While experts couldn't guarantee that it was impossible for humans to contract BSE, their advice was that it was extremely unlikely and had never been seen before. After all, humans had been eating lamb infected with scrapie, the equivalent of BSE in sheep, for 250 years without any negative health effects. But it is hard to prove a negative, and the government couldn't say that eating infected beef posed no danger at all and so it prevaricated. At the same time, there was widespread media speculation over the possibility that BSE could 'jump the species barrier' to humans and a widely held suspicion that the government wasn't telling the public the truth. The media exerted a huge influence over how the public felt, and this situation created a climate of uncertainty and anxiety. In December 1995, an editorial in *New Scientist* warned that 'politicians (and science journals) should not pretend that we know more than we do. In the end, that does more to heighten anxiety and distrust than remove it.'[19]

The government and the public had very different attitudes towards the potential health risks posed by BSE. The government and its official experts were adamant in their view that BSE posed a very low risk to humans so beef was safe. The public, on the other hand, believed there was a much greater risk and it was more than just exaggerated fears. The public had perhaps intuited something about scientific uncertainty that the government failed to understand. Just because there had been no cases of a disease like BSE appearing in humans didn't mean it wasn't possible. Given how many people had been exposed to infected beef, we might have created the exact conditions needed for it to occur, however unlikely. Several factors would have contributed to the elevated public perception of risk during the crisis. The horrific nature of the potential human disease likely created a 'dread effect' whereby people overlook the small chance of it happening and focus on the terrible outcome instead.

It involved strange things like folded prions that were mysterious and almost impossible to understand. It was fatal, affecting the brain, and there was little anyone could do to control it. Most of the population had been exposed, so the consequences would be catastrophic, and it might not appear for years to come. Finally, it was found in our food, the very stuff we need to survive and something we have an intimate relationship with. And it wasn't just any food, it threatened a symbol of British identity and family life: as the French like to point out, the British are *les rosbifs*.

The public's intuition about the risk was vindicated in 1996, when the government finally admitted that at least ten people had died from a new form of brain disease thought to be linked to BSE and caused by eating infected beef. Public concern turned into hysteria, beef sales reached rock-bottom and British beef products were finally banned by the European Union, closely followed by the rest of the world. The human form of BSE, variant Creutzfeldt-Jakob disease, is a fatal brain disease that causes psychiatric problems, behaviour changes and involuntary movements or spasms. In total, 178 people died from vCJD after eating infected beef. After a long and detailed public inquiry ending in 2000, the Phillips Report on the government's handling of the BSE crisis found that, although the government 'did not lie to the public about BSE' and had good reason to believe the risks were remote, the campaign of reassurance was a mistake that undermined confidence in government announcements about risk for years. The report highlighted that the government was driven by an overriding desire to avoid a health scare and so was resistant to changing its position as new information about a possible link with human disease emerged.[20]

During the early stages of the crisis, in 1988, an official in the UK Ministry of Agriculture noted:

We do not know where this disease came from, we do not know how it is spread, and we do not know whether it can be passed to humans. The last point seems to me the most worrying aspect of the problem. There is no evidence that people can be infected, but we cannot say there is no risk.

Yet for the next eight years, everyone in authority told the public there was no risk, promulgating the establishment mantra 'beef is safe'. As Lord Phillips pointed out, this led to a sense of betrayal among the public once it became clear there was a risk and people were dying from vCJD. The gap between the public's perception and the official position illustrates the importance of communicating uncertainty when it comes to risk. During the BSE crisis, the government's false certainty was misplaced and backfired, failing to alleviate anyone's fears. Sometimes, admitting 'we don't know' is a more effective way to secure the public's trust. It also shows that our response to risk is both complex and confusing, especially when it comes to food. In a crisis, experts and officials need to listen to the public more carefully, not only because it can help improve the handling of a situation, but because sometimes they might be right.

Chapter 8

Two Kinds of Water

How inappropriate to call this planet Earth when
it is clearly Ocean.

— Arthur C. Clarke

Seen from space, Earth is a bright blue ball with 70 per cent of
its surface covered by water. It is an ocean world, a type of planet
that contains a vast amount of water, and it is not alone in the
universe. Of the 4,000 planets that have been discovered outside
of our solar system, NASA believes that more than a quarter
could have giant oceans, many of which might be hidden beneath
layers of surface ice. When the *Voyager 1* spacecraft looked back
towards home one last time as it hurtled out of the solar system
in 1990, it took a final series of images of the Earth showing it
as a tiny speck of bluish light, insignificant when seen against
the vastness of space. NASA scientist and author Carl Sagan
described it as a 'pale blue dot', a humble reminder to preserve
and cherish 'the only home we've ever known', and it's the ocean
that makes it home. As if by a miracle, Earth's distance from
the Sun is just right for water to exist as a liquid, being the only
planet positioned in the Goldilocks zone, where it is not too hot
and not too cold – creating the ideal conditions for life to emerge
and thrive. We wouldn't be here today without the ocean, which
gives us life, provides the air that we breathe, feeds us and keeps
the global climate hospitable.

The ocean contains 352 quintillion gallons of water – that's
352 million trillion – and represents 97 per cent of all the water

on the planet (the other 3 per cent being fresh water, most of which is locked up in the ice caps and glaciers). But where did all that water come from? As Joni Mitchell said, 'We are stardust', and so is the ocean. Hydrogen, left over from the Big Bang, and oxygen generated in the cores of stars more massive than the Sun, combine to create water molecules that can be found in huge quantities in the star-forming regions of our galaxy. One of the best-known stellar nurseries, the Orion Nebula, creates enough water every day to fill Earth's oceans sixty times over. When the solar system formed from a swirling disc of gas and dust around 4.6 billion years ago, water would have been abundant. At the time, the infant Earth was too hot, so most of the water that fills our oceans today probably arrived later, after it had cooled down, brought billions of years ago by comets and asteroids in the form of ice and gas. As water accumulated, it formed oceans, lakes and streams. Warmed by the Sun, this water evaporates and rises up into the atmosphere, where it condenses, creating clouds carried by the wind, and eventually falls as rain or snow back to the ground. It then stays there as groundwater or returns to the rivers, streams or lakes, eventually flowing back into the oceans. The same process has been going on for 3.8 billion years – it is the water cycle, and all life on Earth is dependent upon it for its survival. Oceanographer Jacques Cousteau urges us to remember that 'the water cycle and the life cycle are one'.

We need the oceans and our future fate depends on them. They are the lungs of the planet, providing 50 per cent of the oxygen we need to survive. As well as giving us life, they also protect us. They are the largest carbon sink on the planet and are incredibly effective at absorbing excess heat and removing the harmful greenhouse gases that raise the temperature of the Earth causing climate change. Our weather patterns and climate are dictated by the oceans, which carry warm water from the equator to the poles and cold water from the poles to the equator, counteracting the uneven distribution of the Sun's rays and regulating temperatures to keep the Earth habitable. The ocean also gives us food, and, through the water cycle, everything we eat is connected to the ocean. Seafood is the primary source of protein for 50 per cent of the world's population, and coastal nations like Iceland, the Maldives, and Sierra Leone rely on fishing for almost all of their nutritional intake. As if that wasn't enough, the oceans also provide jobs and foster economic growth all around the world, providing livelihoods for three billion people, half the population of the entire planet. Indeed, since the sixteenth century, shipping routes have been the arteries of global trade. Today, the maritime industry connects the world's markets, with 80 per cent of international trade in goods being carried by sea (and that percentage is even higher for developing countries). Yet the oceans are also a mystery. Despite their importance for our survival, around 80 per cent remains unexplored or uncharted. Of the one million different species of animal estimated to live beneath the waves (as well as millions of microorganisms), two-thirds are yet to be discovered, and a staggering 2,000 new species are identified every year.

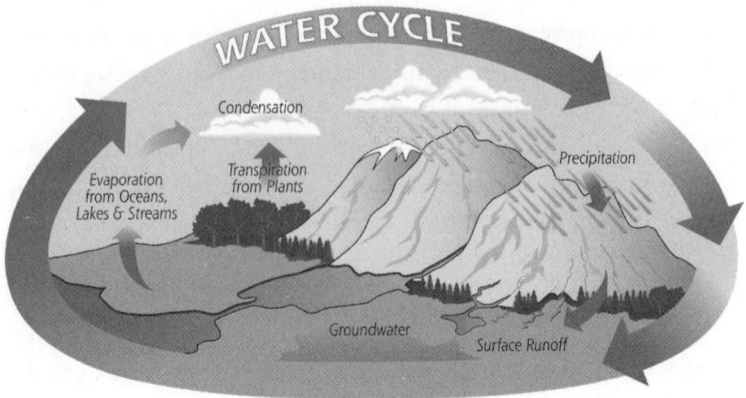

The Water Cycle
(Atmospheric Infrared Sounder, NASA)

Ocean Citizens

For millennia, people have favoured living on the edges of the continent, being within sight of water. Eighty-four per cent of countries have a coastline either with the open ocean or inland seas, and eight of the top ten largest cities in the world are located on the coast, including Tokyo, Mexico, New York, Shanghai and Calcutta. The attractions of living near the sea were clear to the earliest human settlers who created coastal communities over 300,000 years ago that flourished. Where rivers meet the sea, estuaries and alluvial plains provide flat, fertile land perfect for growing crops. The climate is milder, making life easier, and being on the coast gives easy access to the sea or nearby rivers, which is helpful for transport and later became important for trade and commerce. Many early humans settled on the coast not only because of a ready supply of food and other essentials but also because it helped protect them from environmental changes caused by shifts between ice ages.[1] DNA analysis recently revealed that these coastal dwellers might also have been the first pathfinders in the great migration of *Homo sapiens* from Africa to the rest of the world, setting out on a voyage into the unknown in primitive rafts or boats. It's not just our planet that has oceanic origins; our entire species has evolved as an amphibious society spanning land and sea.

At 1,634,701 km, the world's coastlines are so long that if we joined them up, they would stretch around the Earth's equator 402 times. Driven by the need to find work and food or build businesses, or as a lifestyle choice, more people than ever are now living in coastal communities. Around 38 per cent of the world's population live in coastal areas (within 100 km of the sea), and this number is growing rapidly, especially in the developing world, where coastal populations have exploded in recent years.[2] Over the past three decades, the number of people living in coastal zones has increased from 1.6 billion to 2.5 billion, and around three quarters are in developing countries where people have moved to be closer to the economic opportunities booming around ports, fisheries and centres of tourism.[3] One example of this incredible growth is Casablanca, which had just 600 inhabitants in 1839, then 29,000 in 1900, and has just under five million today. Most at risk are the 900 million or 10 per cent of the population who live in low-lying areas (typically less than ten metres above sea level) at the frontlines of climate change, where even a small increase in sea level could cause frequent and devastating floods. The most vulnerable are the poor, and it is estimated that, by 2100, nearly everyone in these low-lying areas will be from low- and middle-income countries.

As more people crowd into coastal cities and towns and the maritime economy takes off, so too does the damage they cause to the environment. Natural landscapes are destroyed or altered to make way for more and more people. Fish stocks, fresh water supplies and land are overexploited, often being irreparably damaged. Coastal waters are drained and land reclaimed, only to be covered in rubbish. Vast quantities of waste and sewage are released into the sea, the volume of pollution increasing exponentially as the population enlarges and industrial activity expands. All this puts pressure on coastal ecosystems such as mangroves, coral reefs and barrier islands, which are fast disappearing. These important natural habitats protect coastal settlements from the worst effects of natural disasters, storm surges and flooding, making them even more susceptible to the impact of climate change and sea-level rises in the future. Ironically, the very things that attract people to the coast in the first place – whether fishing, tourism, natural resources or maritime transport – are slowly being destroyed by the harm caused by the swelling population.

Today, the risks inherent in living near the coast far outweigh the benefits. In its *Global Risks Report 2024*, the World Economic Forum identified sea-level rise as the second-biggest risk facing the world in the coming decade. Melting glaciers, ice sheets and the expansion of seawater as it warms have caused global sea levels to rise by 8–9 inches (21–24 centimetres) since 1880, and most of that increase has happened in the past three decades. The rate at which sea levels are now rising is accelerating, and a new record high was set in 2023. Throughout most of the twentieth century, sea levels rose by an average of 1.4 millimetres per year, but between 2006 and 2015 it rose by 3.6 millimetres annually, almost 2.5 times greater than the average rate over the past 100 years.[4] With Greenland losing 30 million tonnes of ice every hour due to the climate crisis, worst-case scenarios suggest sea levels could rise by up to two metres by the end of the century.[5] A rise of that magnitude would displace, or flood on an annual basis, almost every one of the 900 million people living in low-lying coastal areas. Island nations like the Maldives in the Indian Ocean and Kiribati in the Pacific would be the first to be washed away, while megacities like Jakarta (with a population of ten million), Lagos (population 15.3 million) and Bangkok (population nine million) would be entirely consumed by water. UN Secretary-General António Guterres warned in February 2023 that rising sea levels threaten 'a mass exodus of entire populations on a biblical scale... a death sentence for vulnerable countries'.

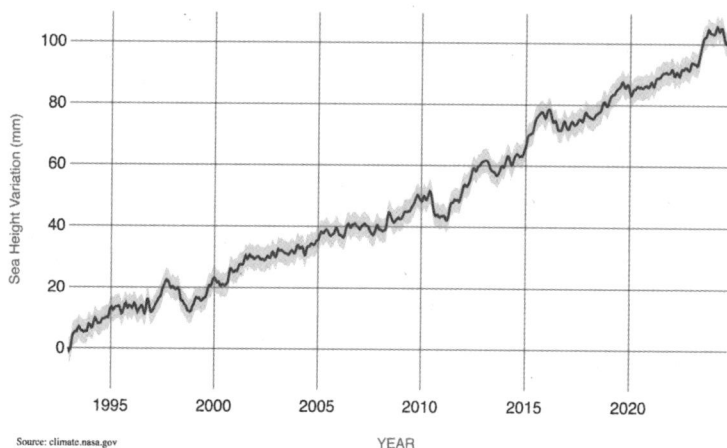

Source: climate.nasa.gov

Rise in sea level since 1993 from satellite sea level observations
(NASA's Goddard Space Flight Centre)

This isn't just a future scenario. It is happening right now and is already damaging lives and wreaking havoc amongst coastal communities. In the Pacific, five islands have already been swallowed by rising seas, and more will soon meet the same fate. In 2016, researchers from the University of Queensland revealed that five of the Solomon Islands, which lie north-east of Australia and south-west of Hawaii, have disappeared without trace. By studying aerial photographs and satellite images taken between 1947 and 2014, they found that five vegetated reef islands had vanished during the sixty-seven-year period. They also found that the shorelines of six more islands are quickly receding. Two villages that had existed since 1935 have already disappeared, and their residents forced to relocate to higher ground on neighbouring volcanic islands.[6] Lead scientist Dr Simon Albert told *New Scientist*, 'It's a perfect storm. There's the background level of global sea-level rise, and then the added pressure of a natural trade wind cycle that has been physically pushing water into the Western Pacific.' He claims their study is the first that scientifically 'confirms the numerous anecdotal accounts from across the Pacific of the dramatic impacts of climate change on coastlines and people'.[7] The Solomon Islands are a stark warning of what might happen to island dwellers all across the Pacific – entire nations could be facing extinction.

Ten miles north of the capital city of the Philippines, groundwater pumping to provide metropolitan Manila with a ready supply of drinking water has caused the land to sink by around two inches in the past twenty years. As sea levels rise, the villages of Bulacan Province, at the north end of Manila Bay, are flooding every time the tide comes in. Many of the older houses are already uninhabitable, and new homes have been built on stilts to keep them above the water level. Locals are having to adapt to having their homes and streets inundated with water on a daily basis. Motorcycles with sidecars ferry people over the flooded streets, and every house has a boat tethered alongside as a means of escape in case the waters rise too high. Climate change is also affecting weather patterns, making severe storms and intense rainfall more frequent. At the end of the nineteenth century, the coastline around Manila Bay was protected by sprawling mangrove forests that have all but disappeared due to urban expansion and the development of fish farming. A unique habitat that provided a home for thousands of species, the mangroves also helped prevent coastal erosion and protected the shore from the worst impact of waves and tidal surges. On average, eight or nine tropical storms make landfall in this part of the world every year, killing thousands and tearing down homes. The destruction of the mangrove forests has left the country's coastal towns defenceless, and it is not surprising that *Time* magazine describes the Philippines as 'the most exposed country in the world to tropical storms'.[8] The ocean itself could be the key to solving the climate crisis. Analysis by the High-Level Panel for a Sustainable Ocean Economy, a coalition of world leaders, shows that the ocean could deliver up to 35 per cent of the annual greenhouse gas emission cuts that need to be made by 2050 if we are to limit the global temperature rise to 1.5 degrees, the threshold that needs to be achieved to avert the worst effects of climate change. Ocean-based renewable energy in the form of offshore wind and floating solar or tidal power generation has the potential to cut carbon emissions by 3.6 gigatonnes per year by 2050, which is more than the current total emissions produced by the whole of India. A further two gigatonnes per year could be reduced if the world's shipping fleets adopt new zero-emission fuels such as hydrogen and ammonia, which would have roughly the same effect as taking 400 million cars off the road every year. Replacing and restoring ocean eco-systems such as mangrove forests will help remove five times more carbon from the atmosphere than tropical forests. Giving up meat and adopting a 'blue food' diet could feed the planet sustainably while slashing emissions even further. It is even possible to capture carbon from the atmosphere and store it safely in the seabed as rocks. The ocean might be a victim of climate change, but it is also a source of potential solutions.

Proteus, the prophetic old man of the sea in Greek mythology, embodied the idea that the seas were highly variable and capable of sudden transformations. Able to see into the future, he might have foreseen that human existence was closely linked to the oceans and that our survival would depend upon our ability to adapt as they change. The obvious planetary pressures of population growth, urban overcrowding and dwindling resources have led many futurists to speculate that our descendants might need to colonize outer space or find a way to live in or on the oceans. Scientists, engineers and science fiction writers have all imagined an aquatic future where humans live in giant floating cities or in special pods under the sea. In 1967, the American architect Buckminster Fuller designed a floating offshore residential structure called Triton City that was capable of housing over 5,000 tenants. Fuller claimed it would be 'resistant to tsunamis' and 'desalinate the very water it would float in'. Fuller was an early pioneer of sustainable urban design, and it captured the imagination of US President Lyndon B. Johnson, who claimed the 3D models for his presidential library. Originally intended to be located in Tokyo Bay, it very nearly became a reality. The City of Baltimore petitioned to have it built on Chesapeake Bay before bureaucratic complications forced Fuller to give up on it. Arthur Conan Doyle's 1929 novel *The Maracot Deep* sees a team of explorers disappear into a deep trench in the Atlantic only to be rescued by the Atlanteans, an empire of submarine people who live on the ocean floor, wearing helmets so they can breathe, foraging for food and mining the seabed for minerals. More recently, the 1995 film *Waterworld* is set in the year 2500 after a climate change apocalypse in which every continent has disappeared under the rising seas, and the whole of human civilization lives on floating communities known as atolls.

Model of Buckminster Fuller's Triton City.

It's not just science fiction. Floating cities are already being built in the Maldives, French Polynesia and Korea. In 2010, billionaire Peter Thiel founded the Seasteading Institute, donating seed funding to build the world's first floating city. In a 2009 essay, Thiel wrote, 'between cyberspace and outer space lies the possibility of settling the oceans'. Seasteading describes itself as a community of 'aquapreneurs' who are working to make floating cities a reality. The UN-sponsored project Oceanix is being built off the coast of South Korea's second-largest city, Busan. When it is completed, three interconnected platforms will provide fifteen acres of space and homes for a community of 12,000 people. One hundred per cent of the energy needed for the city will come from solar cells and it is designed to be self-sufficient, growing its own food, recycling all its waste and replenishing its water supply. In the Indian Ocean, construction is also underway on the Maldives Floating City, which will house 20,000 people and adapt to sea-level rises. The government of the Maldives wanted to turn its country from climate refugees into climate innovators, helping the planet find new solutions to the current crisis. Mohamed Nasheed, president of the Maldives from 2008 to 2012, highlights the importance of working alongside nature: 'Our adaption to climate change mustn't destroy nature, but work with it. In the Maldives we cannot stop the waves, but we can rise with them.'

Jacques Cousteau also believed that the future of humanity was tied to the ocean. In 1962, he gave a speech in London proposing the intentional evolution of the species to live underwater, 'bringing humanity full circle back to the sea'. His vision was to create a race of people called *Homo acquaticus*, using surgery to implant gills in their throats so they could breathe by extracting oxygen directly from the water like fish. Their lungs would be filled with an incompressible liquid so they could withstand the pressures of depth and live deep in the ocean. By 2000, he believed people would be born in seafloor habitats, and, while they would initially need the same surgery as their parents, natural evolution would soon take over. Cousteau's ideas were based on the understanding that the ocean, evolution and human identity are all intertwined. People's hopes and fears relating to rapid environmental change could be addressed by society becoming closer to the oceans. Compelled by climate change, our relationship with the ocean is already changing dramatically and must change if we are to survive. We will all need to be ocean citizens in the future, with or without gills.

A Savage Tigress

Our deep connection to the ocean is apparent in the important role it has played in human culture for thousands of years. It is a common thread running through religion, folklore, ceremonial practice and art all around the world, especially in maritime communities, where it also contributes to a sense of identity. Every year, Hindus on the island of Bali carry small statues of their gods and ancestors from their homes and temples to the sea for a washing ceremony called Melasti. Beautifully attired, they parade to the beach to the beat of traditional music and dip their idols into the sea to cleanse them and recharge their supernatural powers. During the Jewish New Year, Rosh Hashanah, Jews gather by the ocean (or any flowing body of water) to symbolically cast off their sins into the water. In Christianity, water is the primary symbol of baptism; Jesus is 'living water', a symbol of divine life, new birth, growth and the cleansing of sins. In the Bible, water is usually good, but the sea or the ocean represents a barrier from God and symbolizes death or the unknown. The sea is often used as a metaphor for the forces of evil and chaos that threaten to overwhelm humanity. Psalms 89:9 depicts God as the tamer of the dangerous ocean: 'Thou rulest the raging of the sea: when the waves thereof arise, thou stillest them.' For the people who lived in the ancient Near East at the time the Bible was written, the sea was to be feared: it was capricious, a place of danger and mystery where the gods held sway.

The sea also plays a part in many traditional ceremonies relating to the beginning or end of life. It is intimately connected to birth, ageing, death and the concept of immortality. In traditional funeral rites in the Pacific Islands, it was customary to place the dead in a canoe and launch it on the water. Norse chiefs and kings were often buried in ships to allow them safe passage into the afterlife. Scattering the dead's ashes onto water is still common in many parts of Asia, and in India this is usually done on the holy Ganges river. Sailors who died during a long voyage were often wrapped in a shroud made of sailcloth (or a hammock) and committed to the

deep in a ceremony conducted by the ship's chaplain. At the other
end of the human life cycle, the Aoriki of the Solomon Islands initiate their young men during the annual hunt for bonito or skipjack
tuna that appear in giant schools from March to June. Because the
bonito have deep-red blood, they are viewed as manifestations of
the gods and fishing for them is a sacred rite. They fish in canoes far
out in the open sea, a dangerous task that demands great strength.
Therefore, it is both a spiritual awakening and a symbol they are
coming of age and developing the stamina to become adult men. The
sea gives the Aoriki and many other maritime communities a sense
of belonging, an identity that harks back to the way their ancestors
lived, providing a connection to place and a feeling of community.
As E.E. Cummings concludes in the poem 'maggie and milly and
molly and mae', 'It's always ourselves we find in the sea.'⁹

Throughout history, the ocean has been depicted as a hostile
environment populated by fantastic creatures, gods, nymphs,
sirens and monsters. In Homer's *Odyssey*, the sea is a tumultu-
ous 'wine-dark' instrument of the gods that constantly thwarts
Odysseus's attempts to find his way home to Ithaca after being
shipwrecked and lost at sea. Imprisoned on an island for seven
years by the nymph Calypso, he builds a raft and decides to
attempt to sail home. After seventeen days of making good
progress across the Mediterranean, his luck runs out, and he is
spotted by Poseidon, the god of the sea, who angrily whips up
a storm with his trident:

> Poseidon gathered the clouds together and churned up the waves,
> with both hands on his trident. He whipped up
> all the gales from every direction and covered
> both earth and sea with clouds. Night sprang up in the sky.
> Together the East and South Winds clashed, and the raging West
> and North Winds, sprung from the heavens, rolled a huge wave.¹⁰

Homer gives one of the longest (and possibly first) descriptions
of a storm and shipwreck in classical literature. Odysseus falls off

the makeshift boat, which is carried by the winds 'this way and that on the sea'. He tries to swim for it and nearly drowns, believing himself fated to die a 'wretched death'. After two days floating at sea, he sights land, but the danger is not yet over, and he nearly dies after being driven onto a rough, rocky coast by a wave. With a little help from the goddess Athena, he eventually makes it to a smooth beach, 'breathless and speechless', where he is overwhelmed by fatigue but alive. For Odysseus, the sea symbolizes the challenges he must overcome on his journey through life, but it also reveals something about how people in ancient Greece perceived the risks and adversity associated with sea voyages.

Theodoor van Thulden, *Neptune Raises a Storm*
(1632–33, Rijksmuseum via Wikimedia Commons)

The peacefulness and beauty of the sea is a recurring theme in art from all cultures, often juxtaposed with its fearsome destructive potential. Perhaps the best example is Hokusai's *The Great Wave off Kanagawa*, 'possibly the most reproduced image in the history of all art'. Created in 1831, it is an example of the Ukiyo-e woodblock prints or paintings that became popular during the hedonistic Edo period in Japan and makes use of the rich tonal range made possible thanks to the newly invented synthetic pigment Prussian Blue. It shows three boats moving through a stormy sea, almost being engulfed by the giant wave that dominates the foreground, most likely a 'rogue wave' or tsunami. Japanese art became hugely popular in the West after the country reopened following the Meiji Restoration. *Japonisme* was a source of inspiration for Western artists, in particular the impressionists who admired the innovative line style and colours that characterized the Ukiyo-e prints. Van Gogh was struck by the 'terrifying' emotional impact of *The Great Wave*: 'These waves are *claws*, the boat is caught in them, you can feel it.'[11] The composer Claude Debussy kept a copy of the print in his studio and insisted the image be used on the cover of the original 1905 score for *La Mer*, three symphonic sketches that 'run the cosmic gamut from serene tranquillity to terrifying, awe-inspiring power'.[12]

The risks inherent in travelling across the ocean offer a good setting for literature and poetry, full of the threats and the tension needed for good drama. The sea is the primary symbol in Herman Melville's great American novel *Moby Dick*, described by D.H. Lawrence as 'the greatest book of the sea ever written'. Storms, sharks, whales and icebergs all reveal 'that intangible malignity' of the ocean, which 'swallows up ships and crews' like a 'savage tigress' whose 'devilish brilliance and beauty' destroys all.[13] Characters are often seen pitted against the vagaries and hazards of the sea as if in battle against an unknowable foe. However, in *The Old Man and the Sea,* Hemingway's ageing fisherman lives in harmony with the ocean, adapting to its moods instead of fighting against them:

Some of the younger fishermen … spoke of her as a contestant or a place or even an enemy. But the old man always thought of her as feminine and as something that gave or withheld great favours, and if she did wild or wicked things it was because she could not help them.[14]

In his memoirs, the first man to sail single-handed around the world, Captain Joshua Slocum, recognized the poet's attraction to the wildness of the ocean: 'I once knew a writer who, after saying beautiful things about the sea, passed through a Pacific hurricane, and he became a changed man. But where, after all, would be the poetry of the sea were there no wild waves?'[15] On a long sea voyage like his, life and death are ever-present, and this is the central theme of *The Rime of the Ancient Mariner* by Samuel Taylor Coleridge, in which superstition and fate are pivotal. Seafarers are often portrayed at the forefront of ocean risks, never knowing if they will survive. In *The Dry Salvages*, T.S. Eliot speaks of the women who have seen their sons or husbands setting out to sea and not returning, exhorting us to:

Also pray for those who were in ships, and
Ended their voyage on the sand, in the sea's lips.[16]

While the ocean is often venerated for the gifts it bestows as well as its beauty, it is also feared. A powerful force symbolising the unknown, it is often seen as an unpredictable element prone to sudden change and a deadly threat that takes lives and causes destruction. This is the contradictory nature of the ocean. Coastal communities must learn to exist with the knowledge that countless people have lost their lives at sea, sometimes in freak accidents or mysterious circumstances which fuel superstition and anxiety. In the documentary film *Two Kinds of Water*, director Dan McDougall exposes the extreme challenges faced by fishermen in Senegal, as they fight to bring food to people's plates in a time of climate change, overfishing and contested waters. The story is told from the perspective of fisherman Ishmaila Mbaye and his wife Koumba – a couple fighting to stay afloat in one of Africa's most vulnerable fishing communities. Due to overfishing by larger trawlers and depleted stocks of the sardines that were their staple catch, the local fishermen have to travel further into the dangerous North Atlantic for octopus. The film clearly shows the couple's mixed feelings about the ocean: they rely upon it to feed their family but live in the knowledge that one day Ishmaila might not come home. The documentary's title is inspired by a passage in the Qur'an which highlights this contradictory nature:

> He is the one who has set free the two kinds of water,
> One sweet and palatable, and the other salty and bitter.

The Most Dangerous Job in the World

While everyone connected to the ocean faces increasing danger as the climate crisis deepens, it is the people who work in or on the sea who face the biggest risks today. Jobs at sea are amongst the deadliest you could have, whether as a fisherman, seafarer or on one of the world's 240 fixed offshore oil rigs. They are ranked alongside more obviously risky professions like being a lumberjack, landmine remover, venom milker, stuntman or ice-road trucker. Whether you work in an office, on a farm or at a construction site, the work environment is a dangerous place for a large proportion of the planet. The 2023 *World Risk Poll* surveyed people in 142 countries and found that 18 per cent of the global workforce had experienced injury or harm at work in the past two years. With 3.5 billion people estimated to be employed worldwide, that's a staggering 630 million people, roughly equivalent to the entire population of Europe. Fishing has the highest levels of harm of any industry, with one in four fishermen saying they have personally experienced harm in the last two years.

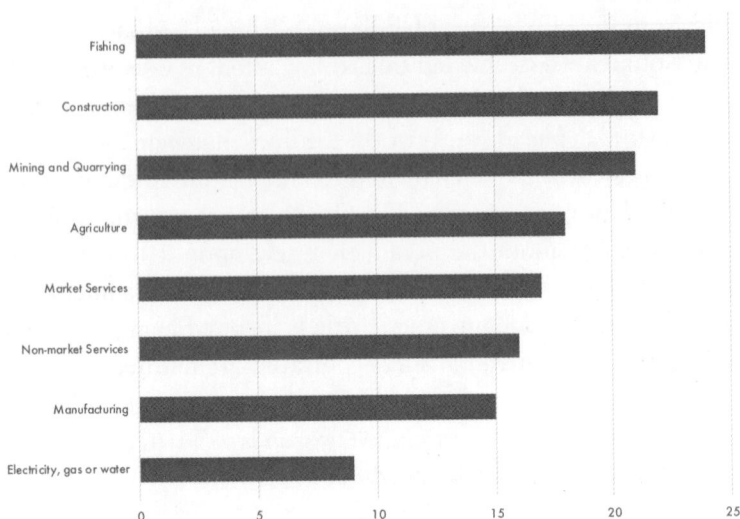

Recent experience of harm at work among the current workforce, by global industry (% who have experienced or know someone who has) (2023 Lloyd's Register Foundation *World Risk Poll* powered by Gallup)

Fishing has always been a dangerous profession, and the people who work on fishing boats work long hours in poor conditions, often battling extreme weather to bring home a catch. Common accidents include falling overboard, capsizing, getting trapped in machinery or fishing gear, fire and collisions with other vessels. As boats pitch and yaw in rough seas, decks become wet and falls or trips are commonplace. These might not always be serious, but when working in remote waters far from land, even a minor accident can quickly become fatal. One of the worst fishing disasters ever to take place in the UK, named 'Black Friday', happened in October 1881 at Eyemouth in Berwickshire on the east coast of Scotland. The only safe landing place on that part of the coastline, Eyemouth had always been associated with the sea and fishing, dating back to the thirteenth century when the Benedictine monks of Coldingham Priory acquired the right to fish in local waters. By the nineteenth century, it was a busy fishing port with a large fleet landing huge catches of haddock and herring. Lying in the path of eastward-moving Atlantic depressions, it is also prone to wild weather and violent storms. The year 1881 was particularly bad, and storms battered the coastline throughout the spring and summer, making fishing impossible and leaving many families hungry. On Friday, 14th October, the sea appeared calm, and, although the pressure remained low, the entire fleet set out to fish for late herring. After sailing for three hours, the fleet arrived at the fishing grounds around 8–9 miles from shore and shot their lines. At midday, the wind shifted direction and 'at the same time broke out with terrific and awfully sudden violence and accompanied with rain, completely obscuring the boats from the view of the people on shore and immediately raising a fearful sea along the coast'.[17] As the storm reached hurricane force, the men tried in vain to pull in their nets and get back to land but were swamped by waves. Wives and children watched helplessly from the shore as boats were smashed to pieces or wrecked on the rocks, killing a total of 189 men and boys.

There are around thirty-nine million fishermen in the world working aboard 4.56 million fishing vessels, 82 per cent of which are small boats under twelve metres in length and are mostly of an open design without decks. The vast majority of these are in Asia and Africa and are usually unregulated, despite the fact they can make up to 99 per cent of the fishing fleet.[18] In the Republic of Guinea, a West African country on the Atlantic coast, their 7,000 artisanal fishermen head out to sea in brightly painted wooden canoes that are open to the elements, have no navigation equipment and very little (if any) safety gear. It is estimated that each year, every fifteenth canoe has an accident and that one in every 200 fishermen dies. Overfishing and changes in ocean currents are forcing these fishermen to stay at sea for longer and to fish further out, where they are often run over by industrial vessels. We don't know exactly how many die every year, but conservative estimates over the past two decades have put the number of fatalities at between 24,000 and 32,000 per year.[19] A more recent study by the FISH Safety Foundation found that there could be as many as 100,000 fishing-related deaths every year globally, highlighting the sad fact that, while fishing is inherently risky, many of these deaths are avoidable.[20] The main causes tend to be unsafe or badly maintained vessels, a lack of safety equipment such as navigation lights or buoyancy aids, young and inexperienced crews who are often not well trained and are unprepared for emergency situations or disasters, and poor knowledge of basic safety precautions or first aid. As the demand for seafood increases, fishing could become even more dangerous unless we address these underlying problems.

Many groups are working on solutions from better regulation to education. International lawmakers have been pushing to improve safety and working conditions for the whole fishing industry – from small artisanal boats to big commercial trawlers – but it has been a painfully slow process. In 1977, the International Maritime Organization's Torremolinos Convention set out specific requirements for the construction of fishing vessels and the equipment they carried on board, but it only applied to larger commercial boats over twenty-four metres in length and was never ratified by enough states to come into force. The International Labour Organization has had more success with the Work in Fishing Convention, which became law in November 2017. Known as Convention 188, it sets out binding requirements to address the main issues with working on board fishing vessels of any size including occupational health and safety, medical care at sea, rest periods and written work agreements. It also covers the design and maintenance of fishing vessels so that crews have decent living conditions on board. According to Johnny Hansen, the chair of the International Transport Workers' Federation Fisheries section, it will 'change the work and living conditions for thousands of fishers, working in one of the most dangerous and often unpoliced professions in the world. Far too many of them are scandalously and criminally exploited. This should be a turning point in their lives.'

To date, Convention 188 is doing slightly better than previous attempts and has been adopted by twenty-two countries, including Angola, Congo, Morocco, Argentina, Liberia, Senegal and Thailand. Ratification is an important commitment, but saving lives relies on how a country implements the convention. This is a complex process that involves consulting the fishing community, changing laws, organising inspections and ensuring there are meaningful penalties for those who do not comply.

Better regulation is a long-term solution and will provoke change but slowly. If we are to keep the number of fatalities from increasing further, we need solutions that will have an impact right now. Fishermen who have seen friends or relatives die at sea don't need to be told that safety is essential. They need reliable engines, basic safety equipment and detailed knowledge to make good decisions, use the equipment effectively, respond to an emergency or provide first aid to someone who has had an accident. The profits from small-scale fishing are meagre at best, so skippers rarely have any money left over to invest in safety equipment. Providing lifejackets, flares, non-slip decking, navigation lights, reliable engines, or a GPS tracker could make a huge difference. But there is no point in providing safety equipment if nobody knows how to use it.

Charities like the FISH Safety Foundation are helping to train the world's poorest fishermen in countries like Bangladesh and the Philippines, giving them the knowledge they need to keep themselves and their crewmates safe. Founder Eric Holliday explains: 'The death rate among fishermen in Bangladesh is more than twice as high as the global average. The toll this takes on communities is unacceptable. Training, and simple interventions such as access to safe drinking water and lifejackets, can make a big difference.' Funding for such practical safety interventions is hard to come by, but a recently created International Fund for Fishing Safety is beginning to plug that hole. Established by Lloyd's Register Foundation and managed by the Seafarers' Charity, it supports safety initiatives led by fisher organizations worldwide and has helped 65,000 fishers in its first year alone.

A study of Vietnamese fishermen found that they are less risk-averse than people in other occupations (farmers, salespeople, government officials) and the unemployed.[21] Among the poorest fishermen, the need to make money and provide for a family is a powerful factor and has a big influence on their risk tolerance, sometimes leading them to fish when it would be safer to stay ashore. US commercial fishermen have been found to consistently underestimate the risks they face at sea, especially if they come from a fishing heritage where the dangers of the job are seen as 'normal life', making them overconfident and apt to ignore safety rules.[22] If fishermen don't believe there is much risk in what they do or don't think they can do anything about it, they are more likely to act recklessly, won't participate in voluntary training, and will propagate a culture that doesn't value safety. While the dangers of their profession are well known amongst fishermen, they have an unusually high risk tolerance, which affects the choices they make, such as where to fish, how far out to go or whether to go to sea at all in bad weather. In many fishing communities, the hazards of the job are such an integral part of daily life that they are seen as trivial or inevitable and so not worth worrying about. Most fishermen have experienced accidents or know someone who has died at sea, giving them a fatalistic attitude towards the risks. Fishermen also tend to be more 'adventurous' than workers in land-based jobs, sometimes exhibiting risk-loving personalities. So, if we are to successfully reduce the number of fishermen who die at sea every year, we must also understand the complex ways that local culture and personal circumstances influence their perception of risk and the decisions they make.

The Spice Islander

Late at night on 9th September 2011, the MV *Spice Islander* left the port of Unguja in Zanzibar's largest island, heavily laden with cargo and jam-packed with passengers. It was travelling north through the Zanzibar archipelago to the island of Pemba, a popular tourist destination and one of the best scuba-diving resorts in the world. The ship's official capacity was forty-five crew and 645 passengers, but it was later discovered that they were carrying somewhere between 2,000 and 3,000 passengers (the exact number remains unknown). The ferry was built forty-four years earlier and was originally a Greek ship called *Marianna* that ran between the mainland and the Greek islands until it was sold to a Honduran company and put into service in Zanzibar as the *Spice Islander*. Rusty, falling apart and badly maintained, she had already experienced engine trouble in 2007 when sailing off the coast of Somalia and had to be rescued by the US Navy. Passengers said they knew something was wrong with the ferry from the outset. Aze Faki Chande is a twenty-seven-year-old mother who spent seven hours at sea and lost her two children and sister in the accident. Lying on a mattress among survivors crowded into the Mnazi Mmoja Hospital in Stone Town, she told Reuters: 'The ferry was clearly faulty even before we started the journey at the Zanzibar port on Friday night. It was leaning to one side. A few of the passengers managed to get off the ship after noticing that it was tilting. We also tried to disembark, but the ship's crew quickly removed the ladder and started sailing toward Pemba.' Local residents were angry that the vessel had been allowed to leave port and said they had frequently referred to the ferry as a disaster waiting to happen.[23]

At around one o'clock in the morning, as passengers tried to sleep in the cramped conditions, the ship suddenly lost engine power. Being so full of passengers made it top-heavy so it quickly capsized, turning people upside down and pitching thousands of them into the deep sea in the middle of the night. Those lucky enough to find something to cling to – mattresses, tables, crates, whatever floated – drifted on the dark waters for hours until the strong currents washed them up on the sandy shores of Zanzibar. Fifteen-year-old Yahya Hussein, who floated to safety by grabbing hold of a piece of wood with three others, told reporters she 'realized something strange in the movement of the ship. It was like zigzag or dizziness. After I noticed that, I jumped to the rear side of the ship, and a few minutes later, the ship went lopsided.'

As the sun rose the next day, thousands of people crowded the beaches, desperately hoping family members would emerge from the waves. Tourists volunteered to help the local authorities, who were overwhelmed as the dead and injured were washed ashore. Policemen waded through the water all day long carrying bodies on stretchers, helping those who made it and talking to family members searching for lost relatives. Photographs in the next day's media showing lifeless children being carried from the sea came to define the tragedy and still haunt the local community. In total, 1,573 people are believed to have perished, although only 240 bodies were ever recovered. There were 620 survivors, forty of whom had serious injuries.

Zanzibar's Minister of State, Mohamed Aboud Mohamed, told a news conference the next day, 'The government will take stern measures against those found responsible for this tragedy, in accordance with the country's laws and regulations. We will not spare anyone.' However, the government failed to take any real action, and the ageing fleet of ferries that the residents of Zanzibar relied upon continued to operate as if nothing had happened. Eight months later, another ferry sailing the same route sank with an estimated loss of 293 lives. For the islanders of Africa and Asia, travelling by ferry is one of their biggest risks, yet something they are forced to do. Most people in Zanzibar can't afford the expensive catamarans that transport tourists during the day, so they have to travel on night ferries, which are old, usually overloaded, and frequently break down. While complaints are often made, the government says there is no money to update the fleet. Munira Ahmed, a resident of Stone Town, Zanzibar, says, 'Government officials do not carry out any routine inspections on these ships to ensure they are safe. We are risking our lives every day in these waters. Authorities look the other way while these ferries pack passengers like sardines and overload the vessels to dangerous levels.'[24] There has been little justice for the families who lost loved ones on board the *Spice Islander*. The captain, owners and director of the Marine Transport Authority were charged with 229 counts of manslaughter for 'failing to take safety precautions' and overloading the vessel. However, two years later, the case was struck off by the Zanzibar High Court due to 'technicalities' and 'defective charges and investigations'. All twelve people who were accused were acquitted and allowed to go free.

The *Spice Islander* wasn't an isolated incident. After fishing, ferries are the second-biggest risk people experience at sea. On average, around 1,000 people die every year in a passenger ferry accident, and the vast majority of these are domestic ferries in Asia and Africa. In 2018, Lloyd's Register Foundation investigated safety on passenger ferries and found that the highest number of fatalities occurred in the Philippines, Bangladesh and Indonesia.[25] Some of the causes are the same as fishing: the absence of regulations or the ability to enforce them, inexperienced and poorly trained crews and a lack of good weather information (or a failure to pay proper attention to weather forecasts). Like the *Spice Islander*, ferries in poorer countries are often old and badly maintained. In fact, many of the ferries currently in use in these regions wouldn't be considered seaworthy if they were in Europe. Overcrowding is also a significant issue and the root cause of roughly one-quarter of the accidents in the Philippines, Bangladesh and Indonesia.

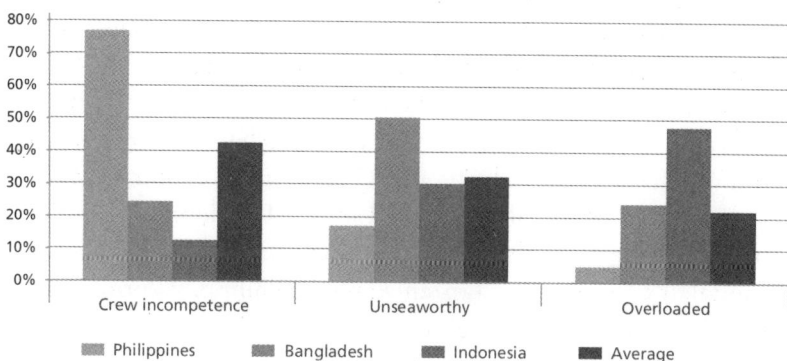

Root causes of ferry fatalities in Bangladesh, Indonesia, and the Philippines (Lloyd's Register Foundation, 2018. Data from Baird, N, Baird Maritime Passenger Accident Database, Baird Publications)

An archipelago of 7,461 islands, the Philippines is totally dependent on ferries. Every day, thousands of people travel on boats shuttling between the islands, crossing volatile seas prone to typhoons. The vessels vary in size from small, motorized outriggers known as *bancas* to roll-on/roll-off car ferries, large passenger ships and state-of-the-art high-speed catamarans. Since 2016, this ocean nation has achieved a remarkable turnaround in their safety statistics – going from the world's worst to close to the global average. Following the publication of its report on ferry safety, Lloyd's Register Foundation established a project called Ferrysafe, in collaboration with the international industry association Interferry, to find out what they were doing that was so successful and help other countries in the region to do the same. Project director Johann Roos explained, 'We needed to find out which of the measures taken have been the most efficient and successful and could potentially be exported to other countries with high levels of ferry fatalities.' The Ferrysafe team interviewed ferry owners, crews and port authorities and found that, while there was no 'silver bullet', there were some critical factors. These included government willingness to change, the introduction of stringent regulations, banning sailing in bad weather, and making sure there were local enforcement officers in ferry terminals.[26] The solutions might appear simple, but for developing countries, the state of their economy often dictates what they can do. As Roos points out, 'We know which regulatory mechanisms are needed to make ferries safer. The problem is that many countries in the region haven't been able to afford them.' Developing countries have to make hard choices about where to invest their limited capital; if countries like Tanzania and the Philippines prioritized the biggest risks facing their citizens every day, then ferry safety would be their number one concern.

Coffin Ships

The first goods to be transported by water were probably simple loads fastened to the back of a large log and floated down a river. Early humans used rafts and primitive boats to transport themselves from one shore to another, and the ancient Egyptians used sailboats made of papyrus to traverse the Nile. It wasn't until maritime trade between different countries and across larger distances started to become more common (and more profitable) that capable seafarers willing to accept the risks inherent in long sea voyages became needed. The Hanseatic League was one of the first maritime trade associations and dominated shipping in the North and Baltic Seas between the thirteenth and fifteenth centuries, trading fur, woollen fabrics, beeswax and grains (among other things) between the cities of Novgorod, Bruges, London and Bergen. It operated along the coastline of what is now Belgium, the Netherlands, Germany, Denmark and Poland, and as far along the Baltic as Estonia. Their single-masted ships, known as 'cogs', were incredibly efficient and able to take a large cargo with just a small crew of around forty sailors. The average cog carried around 100 tonnes, as much as a train of fifty wagons pulled by 200 horses. At its height, the Hanseatic League had hundreds of ships and employed thousands of seafarers, but most of its business took place between neighbouring countries along a single coastline.[27]

The era of modern shipping didn't begin until the fifteenth century, as advances in shipbuilding and navigation paved the way for the great voyages of discovery. The establishment of a new sea route to India in 1498 by Vasco da Gama, the European colonization of the Americas and the exploration of the Atlantic and Indian Oceans changed our view of the world. New lands were conquered, trading posts established and a complex network of shipping routes began to take shape. These were the foundations of the interconnected global economy of the twenty-first century. During the Age of Sail, from the sixteenth to the nineteenth century, the shipping industry exploded, and ships were sailing all over the globe. Crews would be at sea for weeks, months or even years, and ships became larger so they could carry more cargo. The East Indiaman, the most advanced ship of its time, was designed to travel the long distance between Europe and India or Southeast Asia. The largest merchant ships to be built in the eighteenth century, they were around fifty metres long, had a capacity of up to 1,400 cubic tonnes and were equipped with twenty-five to thirty cannons to protect them from pirates. They had a crew of around 150 souls who lived in appalling conditions, were usually malnourished and frequently succumbed to tropical diseases. It is estimated that around ten per cent of the crew died on every journey to the east.

Then came the sleek American clippers, which were longer and could carry more sail so they could outpace their European rivals. By 1850, the journey time from England to South Asia was halved; the clipper *Oriental* could sail from Hong Kong to London in just ninety-seven days. Instead of wood, ships began to be built from iron and steel, allowing engineers to push their capacity further, making them bigger and faster. Sails were replaced with steam engines and propellers. Speed and size was the equation that dictated profits in this burgeoning industry, and those who could deliver goods or people quicker and more cheaply stole a march on their competitors. Bigger ships didn't necessarily mean larger crews. Clippers usually had crews of between twenty-five and fifty sailors, significantly less than the East Indiaman. In fact, improved technology meant that ships could be run by much smaller crews, reducing the overheads. Today's diesel-powered megaships measure up to 400 metres in length and have roughly seventeen times the capacity of pre-Second World War freighters but can be operated by just twenty to thirty people. They work on a fleet of almost 6,000 container ships transporting billions of tonnes of freight every year loaded into 226 million standard-size containers. These neatly stacked multicoloured boxes maximize capacity and help streamline the loading and unloading of cargo in ports, making them the most cost-effective ships to date; you can ship a washing machine from China to northern Europe for just $10. As economist Marc Levinson has written, the container ship 'made the world smaller and the world's economy bigger'.[28] The growing efficiency of shipping as a means of transport has made it the fabric of international trade and dramatically increased the industry's need for experienced crew. There are currently around two million seafarers employed on the world's 50,000 ships, transporting every kind of cargo.

The growth of the shipping industry and the evolution of ship design have primarily been driven by commercial interests, sometimes at the expense of the people who work on board. Despite their importance to the success of the industry, seafarers face numerous risks at sea, and safety, which should be the overriding priority, too often takes second place to profit. It has always been a dangerous profession, and seafarers accept risks that many land-based workers would blench at. Until the nineteenth century, seafarers were most afraid of giant sea monsters, being out of sight of land, and freak weather events such as whirlpools. When all ships were made from wood, a seafarer's greatest fear was fire which was difficult to control and could spread quickly. On 18th November 1874, the 1,200-ton *Cospatrick* caught fire south of the Cape of Good Hope while on a voyage from England to Australia. The fire rapidly grew out of control and eventually sunk. Only three of the 472 people on board survived.

The situation has improved over the past 500 years: instead of thousands, now only a few hundred seafarers globally lose their lives at sea each year, but many of those deaths are still avoidable. The risks have changed too, as well as what it means to feel 'safe' when working at sea. Loneliness, mental health and piracy are now the biggest concerns. A study of British merchant seafarers found that between 1976 and 2002, there were 835 traumatic work-related deaths, 564 of which were accidents. Shockingly, fifty-five of these deaths were by suicide. During this period, the frequency of fatal accidents among British seafarers was 27.8 times higher than in the general workforce, making it the most dangerous profession in Great Britain.[29] The same is true in most other countries and the only occupation with a higher fatality rate is fishing.

During the Covid-19 pandemic, it was seafarers who kept supermarket shelves around the world stocked and whose sacrifice enabled the transport of essential food, vaccines and fuel even during the strictest lockdowns. Ships moved and ports stayed open, keeping the world going during the worst global crisis since the Second World War. While many frontline workers were hailed as heroes, seafarers were largely forgotten, and it took a plea from the UN General Assembly, citing a 'humanitarian and safety crisis', before most countries finally designated them as key workers. At the height of the pandemic, some 400,000 seafarers were stranded on ships, unable to return home due to Covid-19 travel restrictions. Many were forced to stay on board long after their contract had ended, some for as long as seventeen months, which made it feel like being in a 'sea prison'. Raphael, a thirty-three-year-old seafarer from the Philippines, was trapped on board a cargo ship for over a year. He told his story to the International Maritime Organization: 'I am tired, exhausted and hopeless. I have been at sea for twelve months already. And we don't know when I can see my kids and family. It's very frustrating. I am trying to show a brave face every day... We deliver the cargo and the goods, but they close the borders for us.' Still doing twelve-hour shifts, Raphael also talked about his fears over safety and the strain it put on the crew's mental health. Hedi Marzougui, an American captain, also expressed his concerns to the UN agency: 'The longer you stay out there, the more fatigued you get physically. The hours, weeks and months start to add up, you get very tired and you are not as sharp. We also have rights as human beings, we have families of our own. We have a life to get back to. We're not robots, we shouldn't be seen as second-class citizens.' Working on a ship is dangerous at the best of times, and extreme fatigue or depression makes life-threatening accidents much more likely.

In her excellent book on shipping and the port cities of the Arabian Peninsula, *Sinews of War and Trade*, Laleh Khalili reminds us that seafarers are often the victims of the 'complex entanglements' that exist between ship owners and nations or international bodies. In the maritime industry, corporate greed and the endless hunt for profit sometimes leaves people behind. As Khalili puts it, 'the seafarers, the port-workers, the coastal inhabitants, those who live along the routes of trade, human connections and the convivial life of port cities are all considered expendable'.[30]

There is no better example of this than the 'coffin ships' of the eighteenth century and Samuel Plimsoll's bitter campaign to save lives at sea. In the nineteenth century, ships were often old or badly maintained, and a huge number were dangerously unfit to be at sea. Unscrupulous ship owners who cared little for the welfare of their sailors would deliberately send ships to sea disastrously overladen, having insured their vessels for enormous sums. When they inevitably sank in rough weather, hundreds of seafarers lost their lives, but the owners collected a handsome profit. Between 1861 and 1870, there were no fewer than 5,826 shipwrecks off the coast of the British Isles and 8,105 people died. Even the Board of Trade recognized that these deaths were preventable, triggering a public outcry about the scandal. Not all ship owners at the time were heartless capitalists. On Samuel Cunard's transatlantic passenger steamers, safety was paramount and he won the contract in 1840 with the motto 'safety first, profit second'. But most British seamen worked on merchant cargo vessels, which were rarely scrutinized. The cause was famously taken up by coal merchant and British MP Samuel Plimsoll, who fought the maritime establishment – and several powerful ship-owning MPs in Parliament – to pass a law that meant all ships had to have a load-line marked on their hull showing how much it can safely carry. This mark became known as the Plimsoll Line and is now an international standard that saves countless lives at sea.[31]

Vanity Fair illustration of 'The Sailors Champion', Samuel Plimsoll, 1873
(Yale Center for British Art via Wikimedia Commons)

The Priceless Key

The maritime industry is changing how it treats both people and the planet, albeit slowly. In the aftermath of Covid-19, seafarer well-being is now being put centre stage by employers and industry associations worried about the sharp rise in mental health problems and suicide rates. Ship-operating companies are teaming up with healthcare professionals to train officers on mental health at sea, including how to identify the tell-tale signs of distress that could lead to suicide. New ships are being designed that put wellbeing first, giving seafarers improved accommodation, more social and leisure spaces and better connectivity so they can speak to family and friends back home. A study of seafarers' mental health and well-being by the Seafarers International Research Centre at Cardiff University found that internet access made a big difference. So did having a good amount of living space on board to sleep, social-ize and keep fit. Professor Helen Sampson, who leads the Centre, explained that 'a lack of space to socialize in and poorly designed, noisy cabins, were key issues that could cause mental health issues such as depression and anxiety. It's a problem that has only become worse with increasing commercial pressures and competition in the freight shipping sector.'

Shipping, responsible for 3 per cent of all human-generated carbon emissions, also needs to address its impact on global tem-perature rises if we are to mitigate the worst effects of climate change. In July 2023, all 175 member states of the International Maritime Organization agreed there would be a 20–30 per cent reduction in shipping's carbon emissions by 2030 and 70–80 per cent by 2040, putting the global maritime industry on the path to achieving net zero by 2050. It takes time to turn an oil tanker around, and that ambitious target won't be met unless companies in the maritime sector make significant financial investments and put the future of the ocean at the very heart of their business plans. Industry giants like Maersk are investing in alternatives to the heavy fuel oil used in most ships: green methanol made from captured carbon dioxide and clean hydrogen (produced using renewable energy with no emissions). They have already launched eight large ocean-going container ships that run on methanol, and the technology is advancing quickly now that the innovation pipeline sees the industry is serious about a change, and is willing to invest in it. Electric ferries are now being seen in many cities and cargo ships with giant sails are being piloted in a potential return to sail power, which has been described as 'one of the most promising energy sources for the decarbonization of shipping'.

In 2019, Larry Fink, CEO of the world's largest investment fund, with over $6 trillion under management, wrote an open letter declaring that to survive in the future, companies must have a purpose and deliver more than just financial returns. He set out how they should demonstrate leadership on social and political issues: 'purpose is not a mere tagline or marketing campaign; it is a company's fundamental reason for being – what it does every day to create value for its stakeholders'. The *Race to Net Zero* is picking up pace in the maritime sector because social purpose is slowly overtaking profit as its guiding star. Companies that rely on the ocean for their business are beginning to realize that the health of the oceans is critical to their survival and pinning their future success to the principle of a 'sustainable ocean economy'. As Friend of Ocean Action and chairman of Lloyd's Register, Thomas Thune Andersen observes, 'The ocean is fast becoming recognized as a priceless key which is vital in unlocking a decarbonized and resilient world'.[32] As the market is gradually learning, if we are to forestall the imminent threat of climate change and secure a sustainable future for everyone, we must redefine our relationship with the ocean and look to the sea for solutions, learning to live and thrive in an oceanic world. As humans become more dependent on the sea, we will need to better understand all the risks that this new ocean living brings with it, making loss of life at sea a thing of the past. We will also have to learn to accept the dual nature of the ocean as both a life-saving resource and an unpredictable hazard, to keep ourselves safe as the ocean citizens of the future.

Chapter 9

Puffing Devils

Crashes between vehicles and other road accidents became a problem as soon as the very first steam engines on wheels began to appear on streets in Europe and the US in the nineteenth century. Richard Trevithick's 'Puffing Devil' was the first passenger-carrying vehicle powered by steam and made its debut journey on Christmas Eve 1801. The experimental vehicle successfully climbed several hundred yards up a 'stiffish' hill, with people hanging on to it, moving 'like a little bird' and travelling faster than a man could walk. Sadly, while the driver and passengers were in a pub celebrating the event, they forgot about the Puffing Devil, which set fire to a shed and was destroyed. These experimental machines, which operated at very high pressure, could be incredibly dangerous. Just a few years earlier, one of Trevithick's colleagues, William Murdock, was testing an early prototype when a local cleric walked by, saw the contraption and died instantly from fright, believing it to be the 'physical manifestation of Satan'. These fantastic devices were a shock to the public, who were in equal parts terrified and intrigued. By the 1850s, the number of steam vehicles on the roads began to rise as the technology improved and they became more popular methods of transport. Although there were very few vehicles on the roads, accidents were rife, mainly due to the careless behaviour of pedestrians who didn't know how to behave around this new kind of carriage. The public also disliked them because they were liable to scare horse-drawn carts, the dominant mode of transport at the time.

To improve public safety, the British Parliament introduced the Locomotives on Highways Act in 1865. An early attempt at road vehicle legislation, it required all motor vehicles to travel at a maximum speed of 4 mph in the country and to 2 mph in cities. It also mandated that a certain number of crew had to be present when a steam vehicle was moving, and a man carrying a red flag should be walking sixty yards ahead of the vehicle to warn others about the coming danger. After the development of the internal combustion engine, the evolution of the automobile accelerated, with the first modern cars appearing around the start of the twentieth century, as gas and diesel gradually took over from steam power. Car racing on public streets became extremely popular, despite the dangers, and there was a push for greater speed as manufacturers vied to build faster and faster cars. In 1904, Henry Ford achieved a speed of 91.37 mph with the Ford 999. Two years later, the barrier of 124 mph was broken by Fred Mariott driving a steam-powered Stanley Steamer, and in 1909 the Blitzen Benz managed to maintain an average speed of 126 mph, a new world record established by a car with an internal combustion engine. But it was Henry Ford's Model T, produced on an assembly line, that finally brought the car to the masses. Started by a hand crank, these affordable family cars had a modest top speed of 40–45 mph. At the turn of the century, there were around 8,000 passenger cars in America; by the outbreak of the Second World War in 1939, this had grown to 26.2 million.

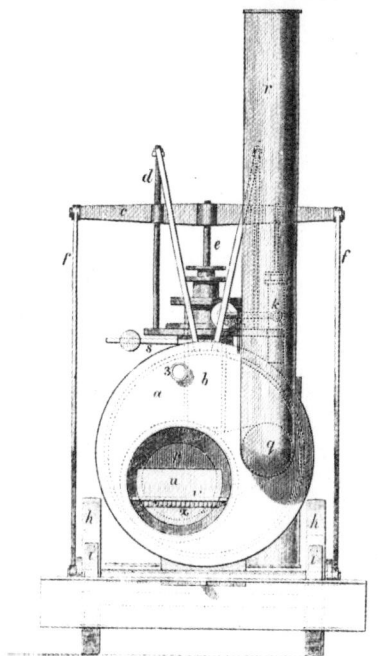

Trevithick's first passenger-carrying common road locomotive, Camborne,
1801 (Francis Trevithick, 1812–77 via Wikimedia Commons)

Number of Deaths

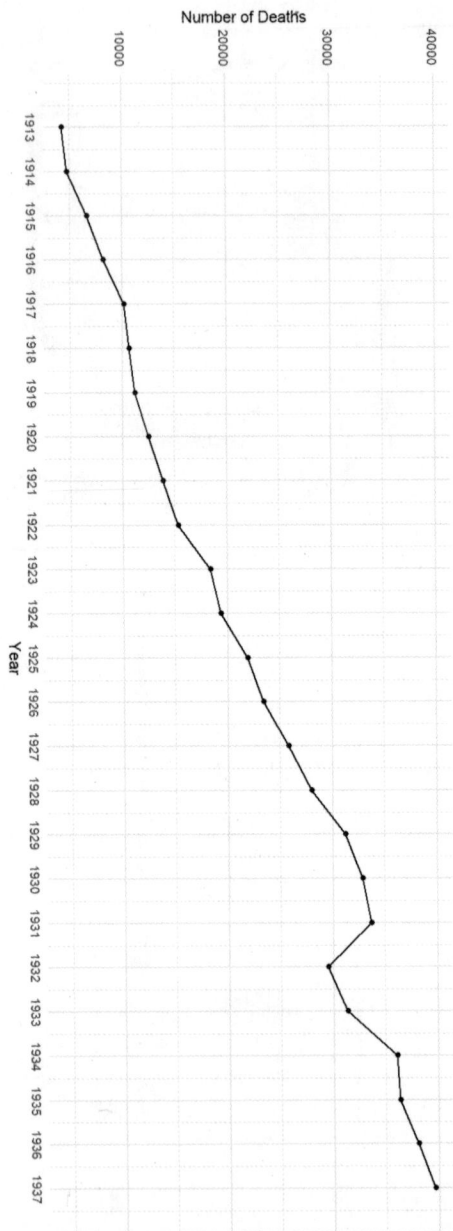

Rapid rise in traffic deaths in USA 1913–17 (data from *US National Center for Health Statistics*. Chart kindly provided by Chia-Wen Wang)

As the number of cars exploded, so too did the number of accidents. Between 1913 and 1937 the number of road traffic deaths in the United States went from 4,200 to 39,643 (an increase of 844 per cent). In the early days of the automobile, safety wasn't seen as an important consideration. Cars were built for comfort, speed, and affordability. Accidents weren't anyone's fault; they were just, well, accidents – something people believed they couldn't control or avoid. For a long time, drivers didn't even need to pass any kind of test to hold a licence. Although France introduced a mandatory driving test as early as 1899, it wasn't until 1935 that a national test was introduced in Great Britain. In America, by 1930 only fifteen states required a driver's exam. As with all new technologies, safety innovation and regulation lagged far behind the development of the main product. However, there were many who campaigned for greater public safety and some notable pioneers. Although he never learned to drive himself, the 'father of traffic safety', William Phelps Eno, created one of the world's first traffic codes for New York City as early as 1903. He was the mastermind behind the stop sign, traffic lights, one-way streets, passing on the right, the taxi stand, pedestrian island and the roundabout (including the traffic circle surrounding the Arc de Triomphe in Paris).

The first recorded case of a pedestrian being killed by a motor car happened on 17th August 1896, when Bridget Driscoll, a forty-four-year-old labourer's wife from Surrey, was visiting Crystal Palace in London accompanied by her daughter and a friend. At the time, there would have been no more than twenty prototype petrol cars in the whole of Great Britain and three new Anglo-French vehicles built by Roger-Benz were being demonstrated nearby. One of the drivers, Arthur Edsall, was taking two passengers on an exhibition ride to experience the curious invention. Victorian onlookers would have been bemused at the sight of the strange 'horseless carriage' which zig-zagged down the road. According to a witness, the car was being driven at a 'tremendous pace, like a fire engine', but in fact, it was only travelling at 4 mph. The driver rang his bell and shouted, 'stand back', but this was likely drowned out by the terrible noise of the engine. Mrs Driscoll was said to have been 'bewildered' and didn't know what to do before she was knocked down. At the inquest, the coroner Percy Morrison said he hoped 'such a thing would never happen again'. Little did he know how much cars would come to dominate our lives over the next one hundred years and how common such accidents would become.

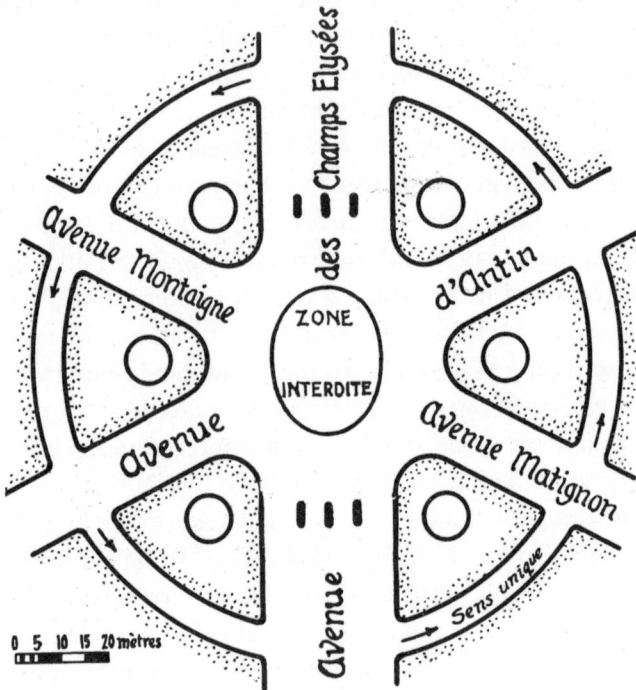

Eno's design for the traffic flow around the Arc de Triomphe, Paris
(Eno Center for Transportation)

Today, more than two people die every minute in a road traffic accident. In 2021 there were approximately 1.2 million deaths on the world's roads, a reduction of around 5 per cent since 2010. It might not sound like much, but it represents significant progress when you consider that the number of cars on the road more than doubled and the global population increased by nearly one billion during the same period. A further twenty to fifty million people are injured every year, many of these resulting in permanent disability.[1] Over the past ten years, road injuries have hovered between the tenth and fourteenth leading cause of death in the world, behind chronic illnesses such as heart disease or lung cancer and communicable diseases such as Covid-19 or tuberculosis. It is by far the leading cause of preventable injury, and it is predicted that it could become the seventh leading cause of death worldwide by 2030. More worryingly, it is the number-one killer of people aged between five and twenty-nine years old. If you are a child or young adult anywhere in the world, the greatest risk in your entire life is being injured in a road traffic accident.

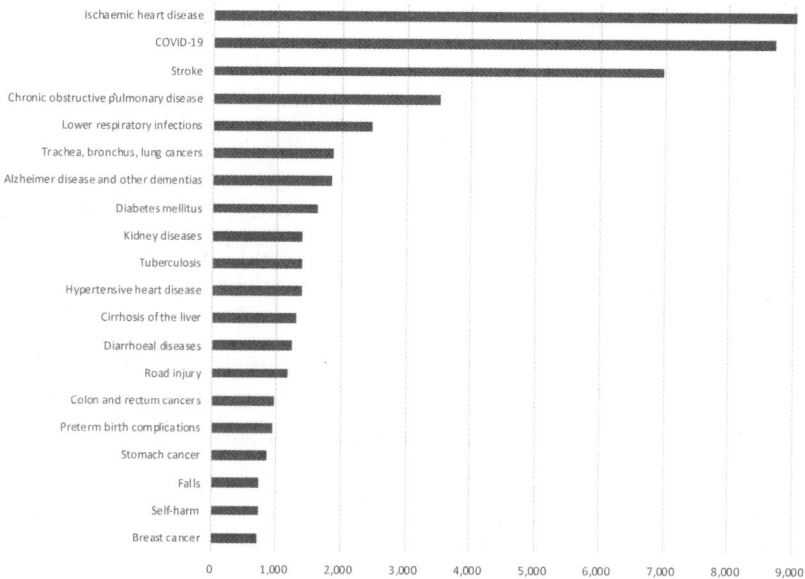

Leading causes of death worldwide, *WHO Global Health Estimates: Leading Causes of Death*, 2021 (thousands)

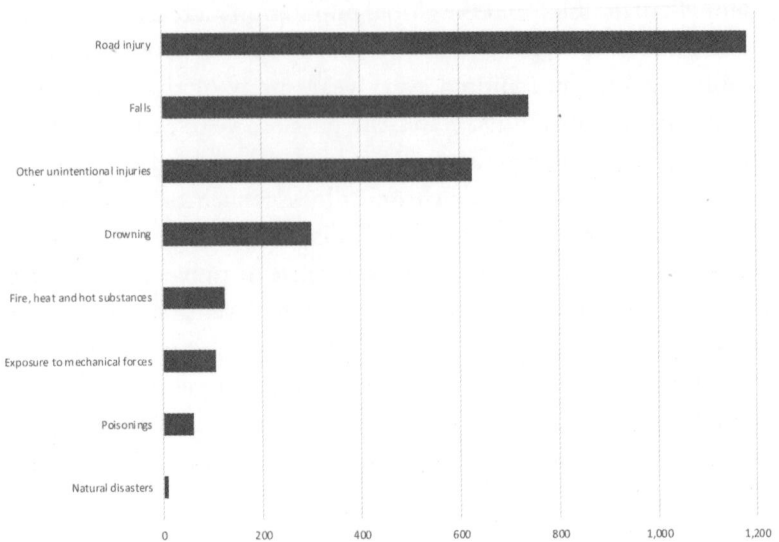

Leading causes of death from preventable injury, *Global Health Estimates: Leading Causes of Death*, 2021, WHO (thousands).

Unsurprisingly, the highest number of accidents occur in countries with the most cars. The United States is by far the most crash-prone, with around two million accidents every year, four times more than any other country in the world. In second place is Japan, with around 500,000 accidents each year, despite being renowned for their safety culture. Germany, the United Kingdom and Canada also feature in the top ten.

Many of these countries have sophisticated road networks, strict safety legislation and a high proportion of cars using the latest technology. So why are there so many accidents? In the United States, the odds of dying in a motor vehicle crash at some point in your life are 1 in 93, only slightly better than the odds of dying after being shot (1 in 89). Natural disasters like storms, dog attacks or aeroplane crashes are all much less likely, although they cause much more worry. For comparison, the odds of dying in a storm are just 1 in 27,925.[2]

Yet, Americans love their cars and feel safe in them. An essential part of daily life, 92 per cent of them have at least one car. It's a necessity when the only way to get to work, take your children to school or to go shopping is by car. But it is much more than that. Since Henry Ford produced the first Model T in 1909, the automobile quickly became an American icon – a symbol of freedom. The classic road trip speaks to what it means to grow up in twentieth-century America, as Jack Kerouac proved with *On the Road*, which came to define the liberation and individuality of the Beat Generation that flourished in the post-war years. Road trip stories are part of the landscape of American mythology, expressing the nation's ideals, and are an important part of the American Dream. However, these narratives don't usually include car crashes or road traffic injuries, an all too likely possibility in real life.

Ann Carlson, chief of the US National Highway Traffic Safety Administration, the body responsible for keeping people safe on America's roads, believes speeding has become a habit, and that many drivers have 'a sense they could get away with it'. In an interview with Reuters in September 2023, she also said that alcohol- and drug-impaired driving remains a significant problem, and that a stubborn 10 per cent of drivers don't wear seatbelts. Carlson pointed out that Americans tolerate thousands of annual road deaths yet still feel safe in their cars, very different to how they feel about aeroplanes: 'If we have one plane crash, it absolutely changes the way people perceive the risk of flying.' American cars are big, luxurious and comfortable; drivers are cocooned in their own private world. Highways are vast, open, well-marked and relatively straight, adding to the feeling of security. All this, combined with the special role cars play in American culture, might well be what causes its citizens to woefully underestimate the risks involved. They are overconfident on the roads and ignore basic precautions. In *Traffic: Why We Drive the Way We Do and What It Says About Us*, Tom Vanderbilt warns, 'When a situation feels dangerous to you, it's probably more safe than you know; when a situation feels safe, that is precisely when you should feel on guard.'[3]

America might be the most crash-prone nation, but it is lucky enough to have excellent emergency services and high-quality medical care available to everyone. If you are involved in an incident in an African country, for example, your chances of survival are much lower. Around 90 per cent of global traffic fatalities occur in low- and middle-income countries. While countries such as Zimbabwe and Vietnam have less than 1 per cent of the world's powered vehicles, they account for 13 per cent of the world's fatalities. The global burden of road traffic injuries is borne by those who can least afford to meet the healthcare and economic costs. Most of the victims in these countries are vulnerable road users, such as motorcyclists or pedestrians, and in many it is a growing public health crisis. Dr Lee Jong-wook, Director-General of the World Health Organization from 2003 to 2006, makes the point that:

> Too often, road safety is treated as a transportation issue, not a public health issue, and road traffic injuries are called 'accidents', although most could be prevented. As a result, many countries put far less effort into understanding and preventing road traffic injuries than they do into preventing diseases that do less harm.

Globally, we are moving in the right direction, but in some of the poorest countries, it is a problem that is deepening. Between 2010 and 2021, 108 countries saw a reduction in the number of deaths (on average of 5 per cent). However, in sixty-six countries there was a rise, and twenty-eight of these were in the African region, which has seen a 17 per cent increase in the number of deaths since 2010.

Emulating many of the symbols of American progress, Africa is now one of the fastest-growing markets for used vehicles, and car ownership is quickly spreading. Yet, across the continent, only 30 per cent of road networks are paved, there are rarely pavements or safe places for pedestrians, road markings are poor or non-existent, and in many places, roads are severely congested, forcing pedestrians, vendors and motorists to fight for space. This is exacerbated by the fact that none of the countries in the African region has national laws that meet best practice for road safety risks.[4] 'Part of the reason for increased fatalities in Africa is the increase in the number of vehicles on the roads,' says Nhan Tran, leader of WHO's safety and mobility unit. 'People who were not able to afford a vehicle ten or twenty years ago can now buy one. Africa has seen a big increase in motorization, but the infrastructure to facilitate it is not there.'[5] Half of all fatalities are among vulnerable road users, meaning pedestrians, cyclists or drivers of two-wheel vehicles. A big increase in the number of motorbikes is partly to blame. Motorbike taxis – called *boda boda*, *piki piki* or *moto* – have become hugely popular in Eastern Africa, a cheap form of public transport that can weave through traffic jams. Motorcyclists have very little protection from an impact, so if a driver hits a motorbike, it is like hitting a pedestrian or worse, especially if they are not wearing a helmet. Some hospitals in the region have even opened dedicated wards for motorcycle accident victims. Elly Kegode, chairman of a motorcycle taxi drivers' cooperative in Kibera, Nairobi, told the *Guardian*: 'I've seen many of my friends die; sometimes their children will come to my office asking for help. It's a sorrow we live with.'[6]

The rapid motorization of the continent has outpaced the development of the infrastructure and regulation needed to keep people safe. Driving while under the influence of alcohol is one of the biggest contributing factors, and only eight African countries have passed national impaired-driving laws that demand a blood alcohol limit of less than 0.05 grams per decilitre, the internationally recommended level. Even in those countries, it is rarely enforced. A 2021 study carried out in Tanzania found that 66 per cent of car drivers involved in accidents were under the influence of alcohol. The other main causes of traffic-related fatalities in Africa include driving when tired, not using seatbelts or wearing helmets, and a propensity to ignore traffic regulations. These are complicated by other issues such as unsafe roads, the lack of available or sufficient post-crash care, inadequate law enforcement and ageing vehicles. But some countries are fighting back. After a series of tragic accidents in January 2023, Senegal announced tough new measures to make their roads safer and are aiming to reduce the number of deaths by at least 50 per cent. In 2021, the government of Côte d'Ivoire decided to enforce helmet-wearing for all cyclists, developed new road safety laws, and created their first traffic police force. There might be a strong commitment to road safety among the region's leaders, but seeing it through will be the hard part, especially given scarce resources and numerous competing priorities. The African region has the highest fatality rate in the world, and solving the crisis will become even more difficult as the race to industrialization tends to prioritize economic progress over public safety.

The main causes of road traffic fatalities have been well-known for many years: speeding, driving while under the influence of alcohol or drugs, not using helmets or seatbelts, distracted driving, unsafe roads and vehicles, lack of post-crash medical care and inadequate laws or law enforcement. Yet, all road traffic injuries can be prevented. In 2021 the UN proclaimed the Decade of Action for Road Safety with the ambitious aim of preventing at least 50 per cent of road traffic deaths and injuries by 2030. This is the second time they have set a decade of action; the first took place between 2011 and 2020 and failed to achieve the target of preventing five million deaths during the ten-year period. The second Decade emphasizes the importance of taking a holistic approach to road safety, highlighting the need to make improvements in the design of roads, manufacturing safer vehicles, enhancing law enforcement and making sure emergency response to accidents is as good as it can be. 'The tragic tally of road crash deaths is heading in the right direction, downwards, but nowhere near fast enough,' says WHO Director-General, Dr Tedros Adhanom Ghebreyesus. 'The carnage on our roads is preventable. We call on all countries to put people rather than cars at the centre of their transport systems, and ensuring the safety of pedestrians, cyclists and other vulnerable road users.'

The Human Heart

If we are to put people at the heart of this system, we must also understand how they think and feel about road safety and, crucially, how those things affect what they do. We know that people's perception of risk has a huge influence on their behaviour, sometimes in unexpected ways. If people believe the probability of experiencing a risk is vanishingly small, then they tend to simply ignore it and get on with their lives. When it comes to the perceived risk involved in driving, the probability of an accident on any given trip is extremely low, and even though we know accidents occasionally happen, we think they always happen to other people. Indeed, every safe journey you have in your car reinforces the idea that nothing will ever go wrong. As the psychologist and pioneer of risk perception research, Paul Slovic points out:

> Such reasoning assumes that people have the unlimited time, energy and attentional capacities needed to have an infinite reservoir of concern. In fact, however, there are only so many things people can worry about and protect themselves against. Unless many hazards are ignored, obsessive preoccupation with risk would preclude any sort of productive life. When choosing which life-threatening events to ignore, those with probabilities near zero are obvious candidates. Indeed, there are many threats that we routinely ignore in order to go on with the business of living: elevators falling, dams bursting, televisions exploding and so forth. For many people, auto accidents may seem so improbable that they fail to incite concern.[7]

When it comes to road safety, research shows that there is often a high degree of rationality and that people's behaviour is directly related to their view of the probability of an accident happening. This has proved a challenge for public safety campaigns that try to shock people into taking road safety more seriously by showing harrowing images of the victims of road traffic accidents and the vehicles themselves. But no matter how horrific people think the consequences will be, driver behaviour doesn't change if they think the likelihood of it happening to them remains low. In other words, successful campaigns for safer behaviour must make people believe the likelihood of the risk itself is high if they are to succeed.[8] We also know that people who are more worried about being harmed (or causing harm) in a traffic accident are often more likely to behave safely – for example, following rules and regulations, wearing a helmet or seatbelt, driving more carefully or purchasing insurance.

Culture also plays an important role. A common English language mistake made by Japanese speakers is 'I am a safety driver' (instead of 'I am a safe driver' or 'I drive safely'), picked up from a popular car commercial and now regularly replicated by English students across the country. Are Japanese people really 'safety drivers'? Opinion is divided, with some believing Japanese drivers to be overly cautious and extremely safe and others adamant that they are bad drivers. In fact, the data shows that accidents are very common in Japan and, as mentioned previously, they are the second-worst country in the world in terms of the overall number of incidents. However, in 2022, the fatality rate had fallen to just 2.6 deaths per 100,000 population, among the lowest in the world. So, while accidents in Japan might be very frequent, they do not normally lead to fatality. This is partly a testament to the efficiency of their emergency response teams and quality of medical care but could also be due to the safety precautions people take. According to the news site *Nippon*, drivers in Japan wear their seatbelt 99 per cent of the time. Whereas American drivers believe they are perfectly safe, research has shown that Japanese drivers tend to overestimate the risk and see themselves as more likely to be at fault. This contributes to them having a much higher sense of 'dread' over road traffic accidents, eliciting visceral feelings of terror, a certainty that any accident will be fatal and the belief that there is little they can do to control the risk. As a result, Japanese drivers buy more car insurance than Americans, and their motivation for this focuses almost exclusively on being able to cover damages or harm done to others and reduce personal worry and stress. It's not surprising in an interdependent society where being ostracized from one's community is a serious penalty.[9]

Since 2019, every two years the *World Risk Poll* has asked people all around the world what they see as the single greatest risk to safety in their daily lives (in their own words). Every single time, road-related accidents have come out as the top answer, even amid major global upheaval during Covid-19, the Ukraine war and the cost-of-living crisis. In 2023, 16 per cent of the world's adult population said road-related accidents were the single greatest risk to safety in their daily lives. It continues to rank higher than the next most-cited risks: crime and violence and personal health conditions.

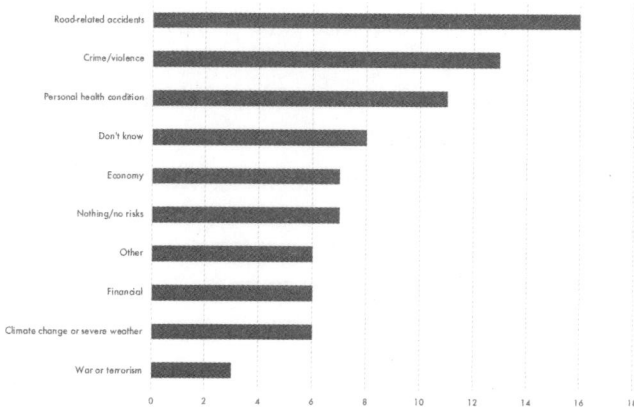

Global top ten risks to safety in daily life – shows percentage who said this was the *greatest* single risk in their daily lives (*World Risk Poll*, 2023)

Traffic accidents also stand out above everything else when you ask people how much they *worry* about specific types of risks. In the 2023 *World Risk Poll*, 76 per cent of adults said they are worried that traffic accidents could cause them serious harm, compared to 71 per cent of people who worry about being harmed by severe weather events and just under two-thirds (65 per cent) who worry about harm from violent crime. The response for traffic accidents was 5 per cent higher than in the 2021 poll, so people worldwide

are becoming more worried about this issue, despite a widespread
reduction in the number of fatalities.

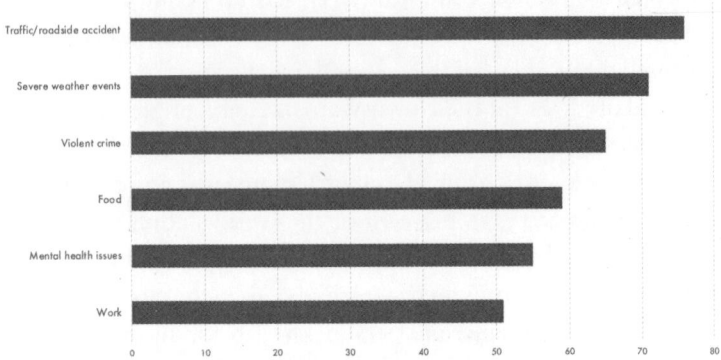

Worry about harm from common risks (*World Risk Poll*, 2023)

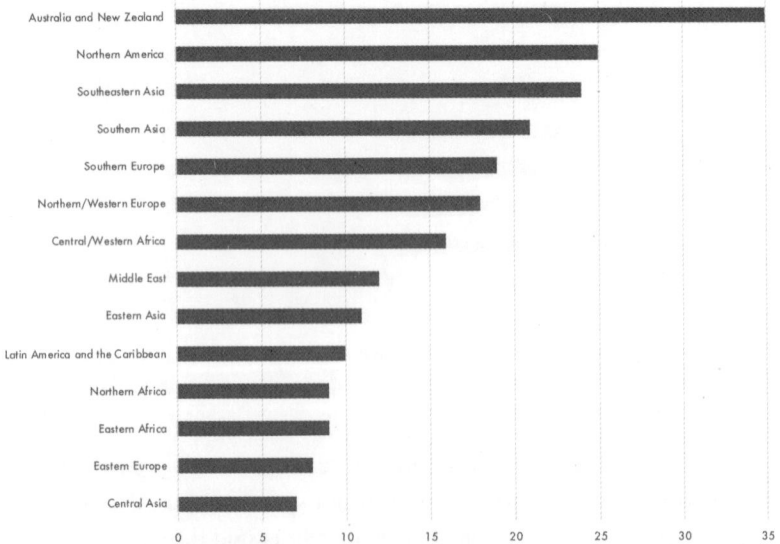

Percentage who said road-related accidents were the greatest risk
to safety in their daily life (*World Risk Poll*, 2023)

People in high-income countries are most likely to name road-
related accidents as the number-one risk to safety in their daily

lives (21 per cent), around twice as likely than those in low-income countries (9 per cent), even though people in high-income countries are far less likely to be injured and have some of the safest roads in the world. In Europe and America, it might seem surprising that road-related accidents are thought of as the *greatest* source of risk in people's daily lives. We know people in those regions are much more likely to die from heart disease or other health-related issues, but these are things people feel they can't control. When it comes to the choices people make every single day, they tend to focus on the things that seem imminent and that they can influence. Parents everywhere are constantly judging the risks they feel might be most likely to cause harm to their family and want to know what they can do to keep their loved ones safe. They also pay attention to the things that scare them the most (so they are easily influenced by the media and high-profile public campaigns). While wealthy countries might have some of the safest roads in the world, it is a risk most people understand and believe they can do something about.

If we look at the African region, the most frequent causes of death reflect the global picture, with road-related injuries being the ninth leading cause of death (slightly higher than the global ranking) behind neonatal conditions, infectious diseases and heart disease. However, people's perceptions about the greatest risk in their daily lives are not necessarily rational and are rarely based on a thorough understanding of both the likelihood and potential severity of all the possible risks. While road-related accidents are still a big risk and no doubt cause widespread worry in Africa, there are probably other things that generate more fear or concern. Many countries in Africa are experiencing rapid urbanization, it hosts some of the most unequal societies in the world, unemployment is high, and it has some of the lowest ratios of police to the public anywhere in the world. These are just some of the factors that cause the region to have a serious crime problem and one of the highest rates of violent crime. Between 1989 and 2023, 3.8 million people worldwide died due to armed conflicts around the world. More than half of these occurred in Africa, where two million people were killed. The year

2023 was one of the most violent since the end of the Cold War. Africa has by far the most state-based conflicts in the world and the number has doubled in the past ten years (since 2013). When you live in such a high-risk environment, it isn't easy to know which risk should be your top priority.

Live Free or Die

While we want people to be safer, *feeling* safe isn't necessarily a good thing. Sometimes, people become more reckless and increase the riskiness of their behaviour in response to improvements in their personal safety. This principle, known as *risk compensation*, has been observed since the early days of the automobile. When the Motor Union of Great Britain suggested people who live next to roads should cut their hedges to make it easier for drivers to see, it produced a surprising result. Colonel Willoughby Verner, an eccentric British soldier, explorer, egg-collecting ornithologist and briefly the professor of topology at the Royal Military Academy Sandhurst, wrote this letter to *The Times* on 13th July 1908:

> Dear Sir,
> Before any of your readers may be induced to cut their hedges as suggested by the secretary of the Motor Union they may like to know my experience of having done so. Four years ago I cut down the hedges and shrubs to a height of 4ft for 30 yards back from the dangerous crossing in this hamlet. The results were twofold: the following summer my garden was smothered with dust caused by fast-driven cars, and the average pace of the passing cars was considerably increased. This was bad enough, but when the culprits secured by the police pleaded that 'it was perfectly safe to go fast' because 'they could see well at the corner', I realised that I had made a mistake. Since then I have let my hedges and shrubs grow, and by planting roses and hops have raised a screen 8ft to 10ft high, by which means the garden is sheltered to some degree from the dust and the speed of many passing cars sensibly

diminished. For it is perfectly plain that there are a large number of motorists who can only be induced to go at a reasonable speed at cross-roads by consideration for their own personal safety.

Hence the advantage to the public of automatically fostering this spirit as I am now doing. To cut hedges is a direct encouragement to reckless driving.

Your obedient servant,
Willoughby Verner.

There are numerous examples of this kind of behaviour. The introduction of parachute rip cords did not reduce the number of sky-diving accidents: it made sky-divers overconfident and they pulled the handle too late. Skydiving engineer and safety pioneer Bill Booth's Rule Number 2 states that: 'The safer skydiving gear becomes, the more chances skydivers will take, in order to keep the fatality rate constant.' Children who wear protective sports equipment tend to play rougher, people who have been vaccinated are more cavalier in exposing themselves to the disease, and levees built to protect against flooding encourage more people to build their homes in dangerous floodplains.

Seatbelts are another example. All cars built in the UK from 1968 had to have front seatbelts fitted but wearing them didn't become compulsory until 1983. America held out much longer, and during the 1980s there was a furious backlash against seat-belt legislation, which the majority of Americans felt represented government overreach in a free society, a slippery slope that would lead to further infringements of civil liberties such as a ban on smoking. One study showed people believed them to be 'ineffec-tive, inconvenient and uncomfortable'. A Gallup poll from 1985 showed that 65 per cent of Americans opposed mandatory seatbelt laws, and many cut the belts out of their cars in protest. But did they work? Things have changed, and now, even in the USA, 90 per cent of people buckle up. New Hampshire is now the only state without a mandatory seatbelt law for adults, but then their state motto is 'live free or die'. According to the US Traffic Safety

Administration, between 1975 and 2017, seatbelts have saved an estimated 374,276 lives and 105 countries now have seatbelt laws. However, many risk experts have pointed out that safety campaigners and policymakers conveniently ignore objective evidence. While it is true that wearing a seatbelt dramatically increases your chances of survival if you are in a car crash, what does it do to the overall number of accidents? In 1975, economist Sam Peltzman published a study of US vehicle safety standards imposed in the 1960s and found that, although they had saved the lives of some vehicle occupants, they had also increased the number of deaths among pedestrians and cyclists.[10] John Adams, a geographer from University College London, published research in 1981 showing that seatbelts led to 'no overall decrease in highway fatalities'.[11] Sometimes the data shows that our behaviour in response to risk is more complex than we might think.

The Safest System

One of the deadliest myths about road traffic accidents is the notion that the vast majority are caused by human error. Every time there is a major incident, news reporters, government officials and law enforcement agencies repeat the widely known statistic that '94 per cent of serious crashes are solely due to human error'. In 2022, the chair of the US National Transportation Safety Board, Jennifer Homendy, called on the federal transportation department to stop using it, saying it was 'misleading' and 'dangerous'. Speaking to the Associated Press, she said the public should be enraged at the number of people who are dying annually in traffic accidents and that promoting the idea that it is all down to driver error creates a culture that accepts it; the idea that it is 'just a risk people take'.[12] Blaming the bad decisions of road users deflects attention from the actions of car manufacturers, road designers and legislators, all of whom share responsibility for improving road safety. The figure stems from a memo on crash statistics published in 2015 which analysed the data from an investigation of 5,470 crashes that had taken place between 2005 and 2007. It found that 'the critical reason, which is the last event in the crash causal chain, was assigned to the driver in 94 per cent of crashes'. However, the memo also pointed out that the 'critical reason' should not be interpreted as the cause of the crash, which might have many other significant factors. In other words, the chain of events that leads to a crash is often complex and the driver is sometimes just the final stage in a long sequence of events. While America still places the responsibility for road safety firmly on the person sitting behind the wheel, Europe has spent the past thirty years taking a very different approach and has seen road deaths fall by over one-third since 2010.

In 1995 a tragic crash occurred on the E4 motorway near Stockholm in Sweden. Five young people were driving too fast on wet roads when they lost control of their car, and it smashed into a concrete lamppost. All five were killed. Claes Tingvall had recently become the head of road safety for the Swedish government, and it was a turning point in how he thought about car crashes. Unlike his predecessors, Tingvall was not a transport engineer or bureaucrat; he had a medical background, having completed a doctorate in the epidemiology of injuries, and so he looked at road safety from a very different perspective. The accident could have been dismissed as simply an inevitable outcome of dangerous driving, but Tingvall saw the injuries as the result of the interaction of a fast-moving vehicle with the built environment and many other variables: road conditions, the behaviour of the driver and the car's safety features. He wanted to understand the entire system and everything that contributed to the fatal injuries that the passengers suffered. For example, the concrete lamp posts were clearly a major problem, and Tingvall went about convincing authorities to remove all such concrete supports on lamp posts next to Sweden's roads, creating the first 'clear zones' along the sides of the country's motorways. It was the beginning of a 'systems approach' to road safety, adopting a safety culture that put the preservation of life above all other considerations, similar to other forms of transport like the aeroplane.

Inspired by his response to the crash, the Swedish minister for infrastructure, Ines Uusmann, asked Tingvall, 'How many deaths should we have as our long-term target in Sweden?' He replied: 'Zero.' Although some believed this target to be idealistic and unworkable, Tingvall felt they had a moral duty to aim for zero. On 22nd May 1997, the Swedish parliament passed a bill that enshrined the target of zero deaths for road fatalities in law. It also made it clear that transport designers were responsible for maintaining the whole system and that it must be based around real people who sometimes made mistakes. As Tingvall explains, 'it should be up to the professional community to make sure that normal people, making normal mistakes, don't lead to them killing themselves or someone else'.[13] Vision Zero is now a global movement and has been adopted by the European Union, Australia, New Zealand, Canada and the United Kingdom, as well as around fifty US cities. In 2020, the UN's Stockholm Declaration was agreed by 140 countries, putting Vision Zero and 'Safe Systems' front and centre in all global policy on road safety. This was also the start of the second Decade of Action for Road Safety, in which the UN, the WHO and other partners call on everyone to implement an 'Integrated Safe Systems' approach based on five elements: safer roads, better transport planning, safer behaviour, safer vehicles and better post-crash response.

The UN's Global Plan for Road Safety acknowledges that speeding, drink-driving, fatigue, distracted driving and not wearing seatbelts, child restraints or helmets are the most common behaviours that lead to road injury and death. However, the actions it suggests countries should implement focus primarily on legislation, enforcement, education and communicating good-quality information to the public. While these are all important, we mustn't forget that human beings are at the centre of the system, and so to make it as safe as it can possibly be, it must be based on a good understanding of how people's attitudes to the risk of road traffic injuries affect their decisions. New measures to improve road safety should be based on good, objective, data as well as insights into the complex psychology that lies behind road users' behaviour. How does their perception of the likelihood of being in a road traffic accident influence their actions? How do people react to feelings of dread or how much they think they can control the risk? What makes people feel safe in a car, and does this have an impact on their response to traffic rules and regulations? We will never reach zero road deaths without a comprehensive understanding of what people think and feel about the risk itself as well as how they might respond, which we know can often be surprising or counterintuitive. The safest system takes account of all the complex factors that influence human behaviour and lead to making good decisions: decisions that make us all safer.

Chapter 10

Surviving the Future

It's dangerous to be alive, and risks are everywhere.
Luckily, not all risks are equally serious.

— Nick Bostrom

The most destructive volcanic eruption in recorded history occurred on 10th April 1815, shrouding the planet in an ash cloud that blocked out sunlight, causing temperatures to plummet and leading to years of disease and famine. After a few days of small-scale eruptions and rumbling, the Tambora volcano in Indonesia suddenly exploded, sending huge plumes of ash skyrocketing into the atmosphere and fast-moving streams of incandescent gas and rock racing down its slopes, wiping out entire villages and killing thousands of local inhabitants instantly. The whole mountain was turned into a flowing mass of 'liquid fire'. When these pyroclastic flows reached the sea, they triggered a tsunami that inundated the surrounding islands. During the eruption, around sixty megatons of sulphur were ejected into the atmosphere, along with rocks and ash, leaving behind a giant caldera 6 km across. A precursor of how modern disasters spread at lightning speed, the Tambora event had terrible repercussions for the whole world that lasted for roughly three years. The planetary chill that ensued caused freak monsoons across the Indian subcontinent that many experts believe gave rise to the cholera epidemic of 1817 that killed millions. It created 'unprecedented snows' in southern China, freezing rain-storms and flash flooding. Rice crops failed, leading to a famine for several seasons. In Yunnan Province, thirty-two-year-old poet

Li Yuyang recorded the events as they happened.[1] His first poem after the eruption begins:

> The clouds like a dragon's breath on the mountains,
> Winds howl, circling and swirling,
> The Rain God shakes the stars, and the rain
> Beats down on the world. An earthquake of rain.
> Water spilling from the eaves deafens me.
> People rush from falling houses in their thousands
> And tens of thousands, for the work of the rain
> Is worse than the work of thieves. Bricks crack. Walls fall.
> In an instant, the house is gone.
>
>   – Li Yuyang, 'A Sigh for Autumn Rain', 1816
>   (trans. Wood, 2014)

The year 1816 became known as 'the year without a summer', and people everywhere struggled to survive due to food shortages throughout the northern hemisphere. In America, widespread frosts continued well into the summer months, resulting in the failure of most crops. Cities in New England saw snowfall as late as 6th June, and the year was long remembered as the year 'eighteen-hundred-and-froze-to-death'. Reverend Thomas Robbins wrote in his diary, 'I presume no person living has known so poor a crop of corn in New England, at this season, as now.' In Britain, poor harvests caused the price of bread to rise so high that ordinary workers could no longer afford to buy it. 'Bread or blood' riots broke out amongst agricultural labourers in East Anglia, who demanded cheaper food so they could live. When asked why they were rioting, one of the protesters, William Dawson of Outwell, said: 'Here I am between Earth and Sky – so help me God. I would sooner lose my life than go home as I am. Bread I want and Bread I will have.' In 1816, the world's population was largely rural and depended on subsistence agriculture. Millions of people starved or begged for food as crops everywhere failed for the next two years. An unexpected natural disaster had generated a planet-wide climate emergency and global food crisis.

Map of the Sanggar peninsula on the island of Sumbawa, Indonesia, and the crater of Tambora. From Heinrich Zollinger's 1847 expedition to the crater, published in 1855 (University of Oxford, Bodleian Library Collection)

Many contemporary artists were moved by the cataclysmic global weather, either in wonder or to fend off boredom during the dismal summer. Particles lingered in the Earth's upper atmosphere, producing remarkable sunsets and vivid red skies that inspired paintings by J.M.W. Turner and Caspar David Friedrich. Scientists have recently used Friedrich's *Woman Before the Rising Sun* (painted between 1818 and 1820) to help calculate the amount of sulphates and ash that were still left in the atmosphere years after Tambora. Percy Bysshe Shelley and Mary Godwin (who became Mary Shelley later that year) were staying with Lord Byron and other friends at his lakeside villa in Geneva. Forced inside by the abysmal weather, they entertained themselves with a competition to see who could write the best horror story. Mary Shelley conjured her Gothic story *Frankenstein* huddled by the fire while Lord Byron penned the apocalyptic poem 'Darkness':

I had a dream, which was not all a dream.
The bright sun was extinguish'd, and the stars
Did wander darkling in the eternal space,
Rayless, and pathless, and the icy earth
Swung blind and blackening in the moonless air;
Morn came and went—and came, and brought no day,
And men forgot their passions in the dread
Of this their desolation...

– Lord Byron, *The Prisoner of Chillon and Other Poems*
(London, 1816)

Such dark literature reflected the eerie skies and frosty conditions Tambora caused; portents of doom that some believe heralded the end of the world. Did people know what was happening, and how worried were they that this could be an end-of-days catastrophe? It wasn't until much later that scientists discovered the connection between the Tambora eruption and the extreme weather events of 1816–17. At the time, people had no idea of the cause and were blissfully unaware of the true global extent of the crisis. As Gillen D'Arcy Wood highlights in *Tambora*, 'The world's residents were oblivious to the volcanic drivings of their fate.'[2] While 1816 was no doubt the most damaging and widespread in recent history, freak weather was not uncommon between the sixteenth and nineteenth centuries as the Little Ice Age produced unusual planetary cooling. During this period, people became accustomed to sudden climactic shifts and the extremes of heat, drought, rain and snow they brought. Glaciers enlarged, destroying entire villages, and rivers in many European cities froze so deeply they could support ice skating and winter festivals. In London, the Great River Thames Frost Fair took place in 1683, with horse and coach races, bull-baiting, and a giant carnival taking place on the ice. In the aftermath of Tambora, most of the world's citizens would have had no idea how close humanity might have been to total annihilation.

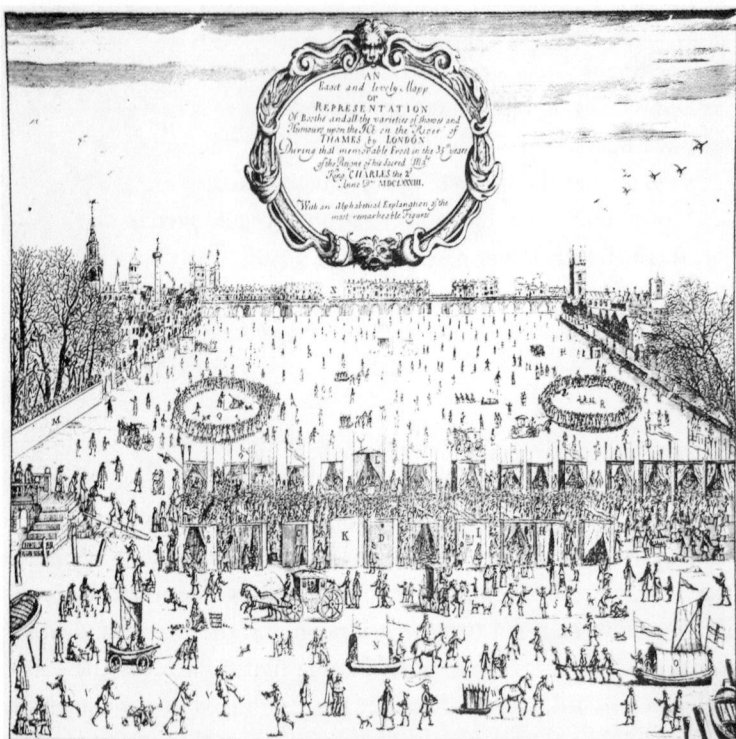

Map showing the 'booths and all the varieties of showes and humours upon the ice on the River of THAMES by LONDON during that memorable Frost' (William Warten, London, 1683)

It could have been much worse. The Tambora eruption is estimated to have been around magnitude 6.9 on the Volcanic Explosivity Index, falling slightly short of being a 'super-volcanic' eruption, a classification that is reserved for magnitudes eight and above. However, any eruption over magnitude seven can cause a volcanic winter lasting decades, forcing a sudden cooling of the Earth, disrupting ocean circulation, and creating a cascade of aftershocks that could decimate global food production. While the planet might endure, it is very possible that all human life could be extinguished in such a global catastrophic event. In 2002, Oxford philosopher Nick Bostrom postulated a new category of risk so severe it could wipe out humanity. He believes these risks demand special attention and preventative action on a planetary scale. He defined an *existential risk* as 'one where an adverse outcome would either annihilate Earth-originating intelligent life or permanently and drastically curtail its potential'.[3] While Tambora was both global and catastrophic, it didn't reach the magnitude of existential; humanity as a whole was able to absorb the shock, survive and return to normal afterwards. However, we might not be so lucky next time, and experts suggest that the likelihood of an existential catastrophe caused by a super-volcanic eruption within the next hundred years could be as high as 0.01 per cent (or 1 in 10,000).[4]

On 24th January 1902, the 'father of science fiction', H.G. Wells, gave a talk at the Royal Institution in London entitled 'The Discovery of the Future', in which he speculated on human destiny.[5] During his lecture, he forecast several improbable but not impossible risks:

One must admit that it is impossible to show why certain things should not utterly destroy and end the entire human race and story... It is conceivable, for example, that some great unexpected mass of matter should presently rush upon us out of space, whirl sun and planets aside like dead leaves before the breeze, and collide with and utterly destroy every spark of life upon this earth... It is conceivable, too, that some pestilence may presently appear, some new disease, that will destroy, not 10 or 15 or 20 per cent. of the earth's inhabitants as pestilences have done in the past, but 100 per cent.; and so end our race... There may arise new animals to prey upon us by land and sea, and there may come some drug or a wrecking madness into the minds of men.

At the end of this statement, Wells confessed that he didn't believe these things would ever happen because of his faith in the 'coherency and purpose in the world'.[6] Natural disasters such as earthquakes, volcanoes and floods are one type of risk with potentially catastrophic consequences. As Wells points out, we also need to consider threats from outer space, disease or pandemics, and the possibility that mankind itself will generate some 'wrecking madness' and destroy itself. Such events have occurred many times in history, yet we have survived. This is reassuring and tells us that such events are incredibly rare and indicates how robust life on Earth can be. The extinction of the dinosaurs, however, is a reminder that even the mightiest of species can fall. The Chicxulub asteroid that hit the planet sixty-six million years ago caused an impact winter lasting fifteen years and wiped out the dinosaurs. More recently, the Tunguska impact that struck the Siberian Taiga in 1908 flattened over two thousand square kilometres of forest.

The Sun also poses a potentially cataclysmic threat. It occasionally throws off massive quantities of magnetized plasma from its upper atmosphere, unleashing massive geomagnetic storms that can cause havoc with the Earth's magnetic field. The largest ever recorded took place in 1859 and is known as the Carrington Event after the British astronomer Richard Carrington, who was busy sketching sunspots at his telescope on 1st September when he noticed a blinding flash of light. A few days later, this coronal mass ejection reached the Earth, causing mayhem with telegraph machines worldwide, which sent sparks raining down on their operators and set papers ablaze. It generated remarkable displays of aurorae so bright that birds began to sing, thinking it was morning, and were visible as far south as the Caribbean and Mexico. *The Times* reported that:

> During the first display the whole of the northern hemisphere was as light as though the sun had set an hour before, and luminous waves rolled up in quick succession as far as the zenith, some a brilliancy sufficient to cast a perceptible shadow on the ground.[7]

In today's electrified world, such an event would cause chaos. Satellites might be fried, sending them off-course and power supplies badly disrupted. A much smaller event in 1989 knocked out Canada's entire power grid, leaving six million people without electricity. A Carrington-scale storm could produce geomagnetically induced currents in the cables that form the backbone of the internet as well as data storage centres, crippling global communications. Physicists put the odds of a geomagnetic storm of this magnitude happening in the next ten years at around one in ten.

Throughout history, human beings have been exposed to a whole landscape of risks ranging from poisonous foods to dangerous animals, epidemics, and war. We have become used to most of these risks and have developed biological and cultural coping mechanisms. Indeed, risk aversion is a behaviour deeply embedded in animal evolution. However, existential risks are new. As Nick Bostrom points out, with the exception of an ancient asteroid impact, 'there were probably no significant existential risks in human history until the mid-twentieth century, and certainly none that it was within our power to do something about'.[8] In weighing up such risks, we need to understand the true scale of their potential impact. In general, risks can be characterized by how many people would be affected (whether personal, local, regional or global) and their likely severity. A third dimension is its probability (how likely the disaster is to occur, given the best available evidence). The diagram below shows that global catastrophic risks occupy the top right quadrant, with existential risks being the most extreme.[9] Using terminology inspired by T.S. Eliot's poem 'The Hollow Men' ('This is the way the world ends / Not with a bang but a whimper'), existential risks can be divided into four categories: Bangs, Crunches, Shrieks and Whimpers. In a 'Bang' scenario, a sudden disaster or act of destruction would cause life to go extinct almost immediately. Some form of life survives in a 'Crunch' or a 'Shriek', while with a 'Whimper' civilization limps on, but is drastically reduced and unrecognizable compared to what existed before.[10] These tools give us a way of understanding and comparing different risks so we can better direct our efforts to prevent them from happening, limit their damage or make it easier to recover afterwards. In a dramatic understatement, Nick Bostrom and Milan Ćirković emphasize that 'our approach to managing such risks must be proactive'.[11]

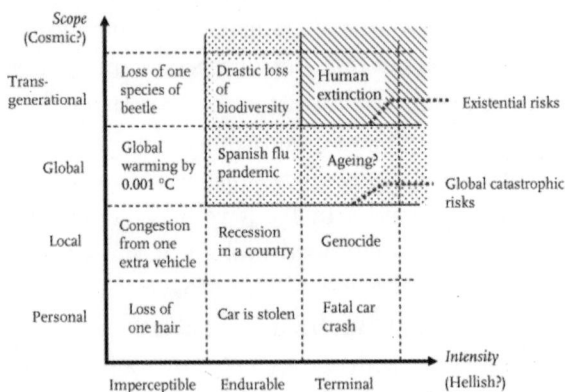

Categories of risk (from Global Catastrophic Risks, OUP 2008)

The risk landscape is changing. In his book *Our Final Century*, Professor Sir Martin Rees writes:

> Throughout most of human history, the worst disasters have been inflicted by environmental forces – floods, earthquakes, volcanoes, and hurricanes – and by pestilence. But the greatest catastrophes of the twentieth century were directly induced by human agency.[12]

The advent of the nuclear bomb was probably the first genuinely existential risk to be created by humans. It raised the possibility that such an explosion could set the atmosphere on fire and vaporize the planet. While it was later proven impossible, many scientists involved in the Manhattan Project, including J. Robert Oppenheimer, believed it to be a 'terrible possibility' and put the likelihood at 'three parts in a million'. Today, the anthropogenic risks that could end the world are familiar to us all: nuclear war, runaway climate change and the threats posed by new technologies such as artificial intelligence or biotechnology. The most significant risks we will face in the future are no longer natural ones; they are being caused by us.

New technologies like artificial intelligence (AI) bring with them enormous benefits and often act as a catalyst for positive social change, lifting people out of poverty and supercharging ailing economies. But every new technology creates new risks. The ability to better understand and manipulate the building blocks of life through synthetic biology promises to improve the environment, raise life expectancy and advance the human condition in extraordinary ways. However, if made widely available, it could also give terrorists the ability to synthesize pathogens more deadly than smallpox. Fears over AI used to be based on the possibility of machine intelligence progressing so rapidly that it becomes self-aware and takes control (like the fictitious Skynet in *The Terminator*). However, a more realistic (and immediate) scenario is that it could give a single country a decisive strategic advantage in a regional or global war. Ironically, AI could also offer protection from existential risks by helping us find solutions or identifying new risks that would otherwise have surprised us. In *The Precipice*, Toby Ord argues that 'the idea that developments in AI may pose an existential risk is not an argument for abandoning AI, but an argument for proceeding with due caution'.[13] In the future, the risks we need to worry most about are likely to be a 'wrecking madness' of our own making, and so we must be vigilant. With existential risks, there are no second chances. Since surviving such an event is impossible, we cannot allow even one to occur. To safeguard humanity, we must understand all the possible threats that new technologies bring with them, even the most speculative ones.

While we are only just beginning to understand existential risks, it is now well-established that we live in an era of global risk.[14] At one time, most of the risks we needed to worry about were local in nature and could be addressed by individuals, communities or national governments. Today, risks are as globalized as our economies, and the biggest threats we face endanger the whole world. Environmental risks such as climate change, biodiversity loss and ecosystem collapse affect us all, while viruses and cyber threats are impossible to contain in our highly interconnected society. In 2006, the World Economic Forum's *Global Risk Report* correctly predicted the danger of a pandemic, describing the 2006 risk landscape as being 'dominated by high impact headline risks, such as terrorism and an influenza pandemic, which top the global risk mitigation agenda and are increasingly well understood'.[15] In its 2024 report, environmental and technological risks dominate predictions for the risk landscape in ten years. According to the WEF survey of global experts, extreme weather, biodiversity loss and shortages of natural resources will be the most severe environmental risks. At the same time, disinformation and the misuse of AI represent the top technological risks (that we can currently foresee).[16] Not only are all these future risks global in scope, but they are also incredibly complex, with multiple contributing factors that are sometimes difficult to disentangle. When such complex systems are involved, solutions are rarely obvious and require commitment from individuals, businesses and nations.

The United Nations Global Assessment Report on Disaster Risk Reduction, *Our World at Risk*, begins with the stark observation that risk creation is currently outstripping risk reduction. Natural disasters, economic loss, poverty and inequality are increasing just as ecosystems and biospheres are at risk of collapse. Its sixth edition, published in 2022, states:

> Global systems are becoming more connected and therefore more vulnerable in an uncertain risk landscape. Covid-19 spread quickly and relentlessly into every corner of the world, and global risks like climate change are having major impacts in every locality.

According to the report, risk is now systemic, driven by our intertwined digital and physical infrastructures, globally integrated supply chains, and increased human mobility. In an age of complex risk, it calls on UN member states and leaders to 'break down siloed thinking and replace it with an all-of-society approach'. But, as it emphasizes, it is not inevitable that risk continues to grow: 'The best defence against systemic risk is to transform systems to make them more resilient.'[17] In the face of such a bewildering array of risks that are so intricately woven together, it is not easy to know what we can do to protect ourselves, our families and our communities. But, as the report states, one thing is clear: resilience is the key to survival both now and in the future, but what exactly is it, and how do we become more resilient?

Bouncing Back

> If you strike a ball sidelong, the rebound will be as much the contrary way; whether there be any such resilience in echoes, that is, whether a man shall hear better if he stand aside the body repercussing, than if he stand where he speaketh, may be tried.
>
> – Sir Francis Bacon, *Sylva Sylvarum* (London, 1625)

'Resilience' means different things to different people but the origin and evolution of the phrase helps us understand what it really means and how it is used today. The derivation of the Latin word *risilio* is unknown, but it is relatively common in ancient Roman literature, usually meaning leaping, jumping or rebounding. In his *Natural History*, Pliny the Elder used the term to describe dolphins and frogs leaping into the air and returning to the sea or land afterwards. Cicero, the greatest orator in the history of the Roman Empire, used it to describe how an accusation can 'rebound back' upon a person.[18] By the Middle Ages, its meaning had altered subtly to mean retract, cancel or 'return to a former position'. One of its earliest uses in English is in the court papers of Henry VIII in 1529, in an exchange of letters about the King's 'great matter' – ending his marriage to his first wife, Catherine of Aragon, so that he could marry Anne Boleyn. At the time,

Catherine was steadfastly refusing to bow to pressure to say that her first marriage had been consummated (which would have allowed Henry to have their marriage annulled). Writing to the Lord Chancellor Cardinal Wolsey, his secretary Stephen Gardiner uses the word *resile* to describe whether Catherine would desist and 'go back' on her word. One hundred years later, its English-language counterpart, *resilience*, was first used scientifically by Sir Francis Bacon, Lord Chancellor of England and one of the greatest thinkers of the Jacobean age. Bacon was the first to formulate the modern scientific method, arguing that the correct way to acquire knowledge of the natural world was through careful observation and inductive reasoning. Putting this philosophy to work in his own scientific investigations, he published a collection of essays on natural history entitled *Sylva Sylvarum* in 1625. In it, he uses the term 'resilience' in a discourse on how echoes bounce off a wall, equating this with how a ball rebounds from a flat surface in different ways depending on the angle at which it is thrown.

After finding its way into English, resilience was never a word that people employed very often. Indeed, it was considered a 'hard word' by contemporary lexicographers and was mainly used by lawyers and philosophers. By the seventeenth century, it had two accepted meanings: to rebound and to go back on one's word. Resilience and its synonym resiliency are included in Samuel Johnson's first authoritative dictionary of the English language, published in 1755. It gives a simple definition of 'the act of starting or leaping back' and cites an example of its use by Bacon in his compendium on natural history. By the nineteenth century, science had well and truly adopted it. It became an essential concept in mechanics and physics – fields that were developing rapidly amidst the backdrop of the Industrial Revolution. In these early definitions of physical resilience, we start to see the origin of the modern concept. Conducting experiments to determine the breaking strength of beams and metal cylinders, English physicist Thomas Young defined resilience in 1807 as the 'ultimate capability of a material to withstand a moving force', for example, a falling weight. Building on Young's work, the railway engineer Thomas Tredgold devised the first 'modulus of resilience [...] which represents the power of a material to resist an impulsive force'. Shortly after, Scottish physicist William Rankine, a leading expert on metal fatigue, defined 'resilience or spring' in 1862 as an 'exact measure of the capacity of a material for resisting shocks by tension'. The way these pioneering material scientists used resilience to describe the strength and flexibility of steel (among other materials) is analogous to its use by civil authorities today in keeping communities safe. As David Alexander explains:

A resilient steel beam survives the application of a force by resisting it with strength (rigidity) and absorbing it with deformation (ductility). By analogy, the strength of a human society under stress is its ability to devise means of resisting disaster and maintaining its integrity (coherence), while the ductility lies in its ability to adapt to circumstances produced by the calamity in order to lessen their impact.[19]

In its simplest form, resilience is, therefore, the capacity of a material or system to return to normal after experiencing stress or shocks. One measure of resilience is how quickly it returns to its usual state. While this focus on returning to the pre-shock state is important for material scientists, it is less useful when thinking about people and the risks they face. When individuals, towns or even whole countries experience a catastrophe, they don't have to return precisely to their 'old normal'. Many academics and researchers prefer to conceptualize resilience as the ability to deal with such shocks and recover quickly, whatever that might look like. As Michael Bruno points out in the *Lloyd's Register Foundation Foresight Review of Resilience Engineering*, 'What matters is preserving and even enhancing the critical functionality of the system, not the pre-existing system'.[20] The international organizations that are most involved in managing significant global risks and responding to disasters have different definitions of resilience. Some, like the Organization for Economic Co-operation and Development (OECD) or European Union, focus on the response, describing it as 'the ability of an individual, a household, a community, a country or a region to withstand, to adapt, and to quickly recover from stressors and shocks'.[21] Others go further, believing true resilience should also include the ability to anticipate and prevent shocks from happening in the first place, or at least having the capacity to absorb them.[22] In *A Farewell to Arms*, Hemingway wrote, 'The world breaks everyone and afterward many are strong at the broken places.'[23] It isn't good enough just to return to the same conditions that existed before a disruptive shock; we need to get better, stronger, and grow. In its most advanced form, resilience means that every bad event or catastrophe transforms the individual or community for the better, boosting their ability to cope with future risks.

What makes us more resilient? The response of the Japanese people to the 2011 Great East Japan Earthquake revealed a wealth of underlying resilience no one knew was there. Realising that the government had different priorities for disaster recovery, residents living in Suetsugi village produced their own maps of radioactive contamination and used them to avoid exposure, while, in other areas, people organized beach cleaning teams in response to 'the needs and wishes of the local community'. They also shared food and other basic provisions with their neighbours. The words of one businessman from Minamisanriku tell what an emotional impact such a relatively small act had:

> There were more than 1,000 cakes in the fridges at the time of the disaster. The next morning, I took them to the rescue centre. When I saw people hurrying to help with the recovery work, carrying a piece of cake with them because they had no time to stop and eat, I felt deeply connected to my community.

New initiatives sprang up across the country, many of which were led by older women who wanted to provide support for themselves and others in the immediate aftermath of the earthquake. In Shichigahama, where 1,000 homes were destroyed, they formed a knitting circle called 'Yarn Alive'. Others started tea clubs, and one group decided to make dolls, telling stories about their childhood and how they survived the tsunami while they worked. Local organizations provided spaces where people could come together and host other activities such as music, art, theatre, and even comedy. All these initiatives have since been shown to have significantly improved community cohesion and mental well-being as people rebuilt their lives following the disaster.[24] A large part of our resilience is based on the strength of such social capital – the human relationships, connections and support that form the invisible fabric of our towns and villages. In a crisis, it is often our neighbours we rely upon the most.

Before the earthquake that devastated Turkey in February 2023, Döndü Karabörk ran a successful glassware business in Kahramanmaraş in south-central Turkey, near the border with Syria. On 6th February, the largest earthquake in Turkish history struck the region, claiming over 62,000 lives and injuring more than 100,000 people. A powerful magnitude 7.8 earthquake was followed later the same day by a magnitude 6.7 aftershock. Around 280,000 buildings were destroyed or severely damaged. Döndü's store and all its delicate stock, together with her family's livelihood, was shattered. Everything she had worked hard to create was lost, and she didn't have enough money to rebuild the store: 'We were about to restart working, but we didn't have capital. I was feeling down.' Döndü was lucky enough to be given a small cash grant from the humanitarian charity Turkish Red Crescent that allowed her to buy the products she needed to reopen her shop and start again. 'That grant of 30,000 Turkish lira (around £675) was very precious to us,' she recalls. 'It was the reason I was able to take my first step back up again.'[25] Her story is far from unique. Thousands of businesses across the region were destroyed, and many families lost their only means of earning income. A large part of resilience is about access to money when it is most needed. If the worst happens, do you have enough money saved up to be able to survive and recover? Can you borrow from friends, family or even neighbours? Can your local community, region or nation provide loans or other kinds of financial assistance in an emergency? This is one reason that disasters have a much more significant impact on the poorest and most vulnerable people and countries. According to the *World Risk Poll*, half of people out of work feel they cannot protect themselves or their families in the event of a disaster.[26] Being in stable employment and having enough savings to withstand an emergency are critical elements of resilience for individuals and households.

Psychological resilience is another vital component. When faced with a crisis or adversity, some people survive and prosper while others struggle. Since the 1990s, psychologists have wanted to understand what makes some people naturally better able to cope with stress, protect themselves or others and often grow personally as a result. The American Psychological Association says it is 'the process of adapting well in the face of adversity, trauma, tragedy, threats or significant sources of stress – such as family and relationship problems, serious health problems, or workplace and financial stressors'. The UK's National Health Service (NHS) defines it more succinctly as 'the ability to maintain personal wellbeing in the face of a challenge.' In the book *Resilience: The Science of Mastering Life's Greatest Challenges*, medical scientists Steven Southwick, Dennis Charney and Jonathan DePierro combine the latest scientific research with the personal experiences of individuals who have survived some of the most traumatic events imaginable. They describe individual resilience as 'the ability to weather and recover from adversity' and believe it is made up of ten factors, including optimism, the ability to face your fears, moral courage and having strong role models. William James, the founding father of psychology in the United States, said, 'Pessimism leads to weakness, optimism to power.' Confirming this idea, Southwick, Charney, and DePierro found that people with incredible personal resilience almost always have an optimistic outlook. They conclude that optimism 'ignites resilience, often providing the energy that drives us to face our challenges head-on. It facilitates an active and creative approach to coping with challenging situations.'[27]

Optimism is one way your mindset can influence how resilient you are. People with a growth (as opposed to a fixed) mindset believe their most basic abilities are malleable and can be improved. This tends to make them more flexible and capable of absorbing, adapting and bouncing back from stressful situations. It has long been known that people with a growth mindset perform better at school and are often high achievers. Companies that encourage (or recruit) employees to have a growth mindset are more productive. This is because people with a growth mindset worry less, seek out challenges and see failure as an opportunity for development. On the other hand, people with a fixed mindset believe their qualities are set in stone, and they can do nothing to change them. In healthcare, growth mindsets can help people eat more healthily and control their weight or overcome addictions such as smoking. For example, a recent study by Carolyn Lo and the team at the Institute for the Public Understanding of Risk at the National University of Singapore found that people with type 2 diabetes tend to have lower growth mindsets than those without the condition, underlining the importance of mindset in making good lifestyle choices and managing your well-being.[28] Cognitive psychologist Carol Dweck explains:

> In a growth mindset, people believe that their most basic abilities can be developed through dedication and hard work – brains and talent are just the starting point. This view creates a love of learning and a resilience that is essential for great accomplishment.[29]

The examples above show different aspects of resilience and all are important, whether financial capital or social capital, individual characteristics or attributes relating to whole communities. Together with good disaster planning and emergency preparedness by national and local authorities, they represent the basic building blocks of a resilient society. They are also connected: social capital has been called the 'core engine of recovery' because it increases both individual and community resilience.[30] These two types of resilience are interrelated. If the people who make up a community have high psychological resilience, the group is more likely to cope well with adversity. Equally, the ability of an individual to respond to shocks depends on the resources available within the community. But none of it works without trust. Communities with high levels of trust have greater social cohesion or solidarity and are more resilient when disaster strikes. Residents in such communities are more likely to help their neighbours, share resources, provide shelter, offer financial support and pass on early warnings or other information.[31] Trust also helps rescue teams to get established and do their job effectively. Trust in experts, institutions and the government is critical in expediting recovery following a disaster. According to political scientist Robert Putnam, trust is the most crucial element of social capital. It is the foundation for cooperation in a community, enabling individuals to work together for mutual benefit. When we trust others, we feel more secure and willing to collaborate with them:

> A society that relies on generalized reciprocity is more efficient than a distrustful society, for the same reason that money is more efficient than barter. Honesty and Trust lubricate the inevitable frictions of social life.[32]

A 'culture of trust' is imperative for a resilient society – both within neighbourhoods and between communities and national institutions or authorities. Strong social ties between residents were found to be one of the central factors in helping people recover from the Kobe earthquake in 1995. However, whatever tenuous trust existed between the public and the Japanese government before the Tōhoku earthquake and Fukushima meltdown in 2011 was so severely damaged that it still hasn't recovered today.

Trust in the government, nuclear power operators and even the media was shattered when a disaster that they had promised could never happen happened. The public was bombarded with muddled and often contradictory information at a time when they desperately needed to make basic decisions about where to live, what to eat and how to look after their children. Suspicion turned to anger as people were left bewildered – not knowing what information to trust or what advice to follow. Local communities were forced to rely on what became known as 'Operation Me' to cope with the disaster's worst effects.[33] As we all know, trust is fragile. It takes a long time to generate, yet it can be lost in seconds. When it is gone, it can take years to rebuild, so trust must be something all groups within a society value and invest in every single day.

The Encourager

> Knowledge is the antidote to fear,—Knowledge, Use and Reason, with its higher aids. The child is as much in danger from a staircase, or the fire-grate, or a bath-tub, or a cat, as the soldier from a cannon or an ambush. Each surmounts the fear as fast as he precisely understands the peril and learns the means of resistance. Each is liable to panic, which is, exactly, the terror of ignorance surrendered to the imagination. Knowledge is the encourager, knowledge that takes fear out of the heart, knowledge and use, which is knowledge in practice.
>
> – Ralph Waldo Emerson, *Society and Solitude*

The risks most of us experience every day involve making decisions – how fast to drive, whether to walk home alone at night or what to do if there is a flood. It is often said that information is the antidote to anxiety, and that is undoubtedly true for everyday risk. Good decisions are based on good knowledge. The more we know about the risks around us, the less we worry and the better armed we are to combat them. Ultimately, we want to protect ourselves, our family and our local community from harm. To do that, we need the kind of risk 'know-how' that goes beyond a basic understanding of what risks exist and their relative probabilities. We also need to know what we can do about those risks, how others in the past have dealt with them, what experts think about the risks, and how different people and cultures have various ways of tackling them. We also need to be aware of the relationships between risks, how they interact, and how they make up the risk landscape that surrounds us.

Understanding risk is fundamentally a human and social endeavour and cannot be reduced to mathematical models that estimate the probability (or exposure) and the potential severity of the consequences. We need to go beyond the numbers and take account of how people think and feel about risk. That involves people's beliefs, attitudes and judgements, which are influenced by their underlying values and cultural perspectives. To understand the risks around us, we must also understand ourselves and the surprising ways our psychological make-up can affect our perception of risk and our behaviour. During the 2011 Tōhoku earthquake, many people lost their lives because they believed it wouldn't affect them, and their lives would continue as normal. This is an example of one kind of bias, known as normalcy bias, in which people underestimate the risk they face. It is a typical human response to overwhelming evidence that a disaster is about to strike. Every time someone survives a disaster, or a disaster wasn't as bad as they were told, it reinforces this idea and makes people even more resistant to acting, especially when it means leaving their homes and belongings. Another human bias, typical in male American drivers, is the tendency to be overly optimistic about good things happening and to believe bad things are much less likely to occur than they really are. Biases like this are often a result of the brain attempting to simplify information to help us make sense of the world around us, but they can be dangerous.[34] They impair our judgement and cause us to make bad decisions. When evaluating risk, confronting and overcoming such biases is critical in protecting people and saving lives.

We all know that 'gut feelings' can sometimes save our lives: risk perceptions are primarily intuitive. They are based on our experience of the world around us, which comes from what people we know tell us, as well as from reporting in the news or on social media. Dangers, unusual events like shark attacks, and significant disasters make the news every day and have a huge influence on our perception of the risks around us. In an attempt at good storytelling or to attract and capture an audience, the media often presents risks dramatically and inaccurately, leading the public to have a massively exaggerated sense of the risk. But what exactly are those perceptions, and which ones have the biggest effect on our decisions? In 1978, researchers from the Decision Research Group in Oregon, USA, were the first to attempt to understand what had the biggest influence on the public's perception of a risk.[35] They studied eighty-one hazards, from nuclear power and pesticides to downhill skiing and terrorism, to find their 'personality'. Their findings, shown in the diagram below, established a new field of research and revealed three major factors that affect our risk perception. The first was labelled 'dread risk' and relates to feelings of terror, catastrophe, something with a fatal outcome, and over which we have no control. Nuclear weapons, nerve gas and terrorism scored highest for this factor. The second was about how 'unknown' the risk is – whether new and mysterious or old and familiar, likely to cause harm now or at some point in the future. Hazards rated highly for this included solar electric power, DNA research and satellite technology, while those rated low included motor vehicles, fire-fighting and mountain climbing. They also identified a third factor mainly related to the number of people exposed. Dread risk is the most important, according to the group's lead researcher, Paul Slovic, because 'the higher a hazard's score on this factor, the higher its perceived risk, the more people want to see its current risks reduced, and they more they want to see strict regulation'.[36]

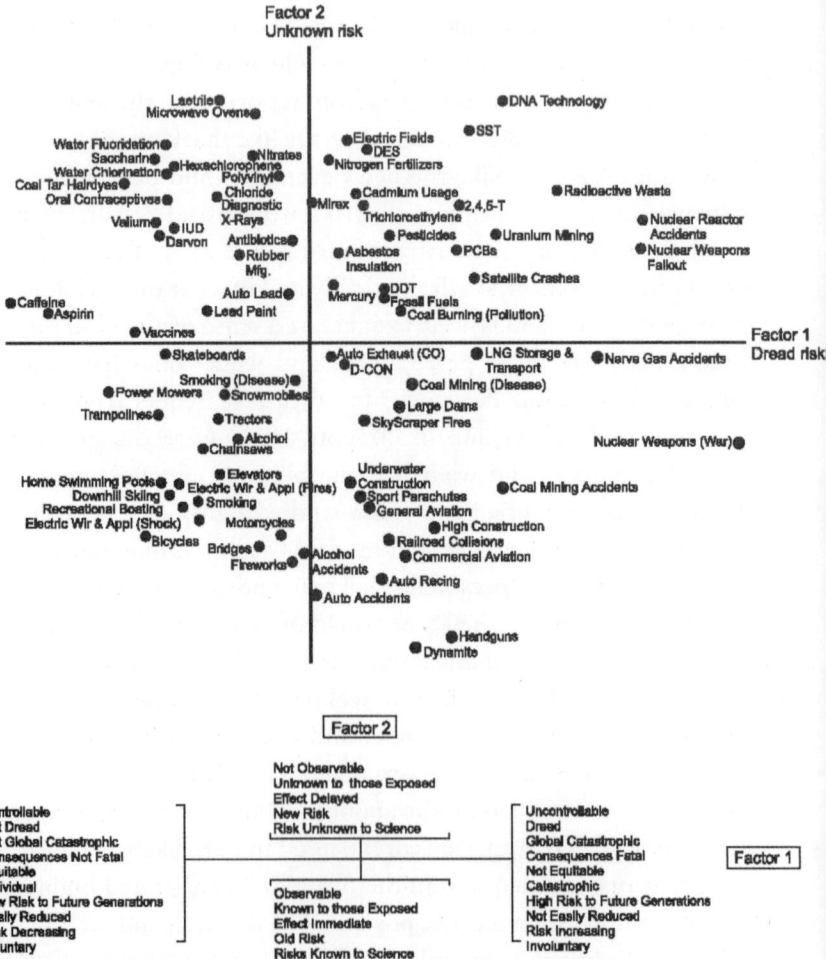

Factor 2
Unknown risk

Laetrile●
Microwave Ovens●

●DNA Technology

●Electric Fields   ●SST
●DES
Nitrogen Fertilizers●

Water Fluoridation●
Saccharin●   ●Nitrates
Water Chlorination● ●Hexachlorophene Polyvinyl●
Coal Tar Hairdyes●   ●Chloride
Oral Contraceptives●   Diagnostic
    Velium●   X-Rays   ●Mirex
        ●IUD   Antibiotics●
        ●Darvon   ●Rubber
            Mfg.

●Cadmium Usage
    ●2,4,5-T   ●Radioactive Waste
Trichloroethylene ●
    ●Pesticides   ●Uranium Mining   ●Nuclear Reactor
    ●Asbestos   ●PCBs   Accidents
    Insulation   ●Nuclear Weapons
        ●Satellite Crashes   Fallout

●Caffeine
●Aspirin
        ●Vaccines

Auto Lead●   ●DDT
    ●Lead Paint   Mercury ●●Fossil Fuels
        ●Coal Burning (Pollution)

Factor 1
Dread risk

●Skateboards   ●Auto Exhaust (CO)   ●LNG Storage &   ●Nerve Gas Accidents
        ●D-CON   Transport
Smoking (Disease)●   ●Coal Mining (Disease)
●Power Mowers ●Snowmobiles
Trampolines●   ●Tractors   ●Large Dams
        ●SkyScraper Fires
        ●Alcohol
    ●Chainsaws   Nuclear Weapons (War)●

    ●Elevators   Underwater
Home Swimming Pools●   ●Construction   ●Coal Mining Accidents
Downhill Skiing ● ●Electric Wir & Appl (Fires) ●Sport Parachutes
Recreational Boating ● ●Smoking   ●General Aviation
Electric Wir & Appl (Shock) ● Motorcycles
    ●Bicycles   ●High Construction
    Bridges ●   ●Railroad Collisions
    Fireworks●   ●Commercial Aviation
        ●Alcohol
        Accidents   ●Auto Racing
            ●Auto Accidents

        ●Handguns
        ●Dynamite

Factor 2

Not Observable
Unknown to those Exposed
Effect Delayed
New Risk
Risk Unknown to Science

Controllable
Not Dread
Not Global Catastrophic
Consequences Not Fatal
Equitable
Individual
Low Risk to Future Generations
Easily Reduced
Risk Decreasing
Voluntary

Uncontrollable
Dread
Global Catastrophic
Consequences Fatal
Not Equitable
Catastrophic
High Risk to Future Generations
Not Easily Reduced
Risk Increasing
Involuntary

Factor 1

Observable
Known to those Exposed
Effect Immediate
Old Risk
Risks Known to Science

Location of 81 different hazards in 'risk space'. The position of each hazard is given by the combination of the characteristics shown in the table at the bottom of the diagram (kindly reproduced with permission of Professor Paul Slovic)

In contrast, experts' perceptions of risk are not related to any of the factors described by the Oregon team. Nuclear scientists, doctors, and engineers tend to have a more objective view of the 'riskiness' of such hazards, largely based on their knowledge of the annual mortality statistics. Slovic rightly pointed out that 'conflicts over risk may result from experts and lay people having different definitions of the concept'.[37] People's risk perceptions are often very different to those of experts and are rarely based on the true statistical likelihood of harm. The gulf between such expert knowledge and public perception has been called the 'risk perception gap' and can prevent us from making choices that keep us healthier, happier and safer. After the 9/11 terrorist attack on the US, many travellers were afraid to fly and chose to drive instead, a much more dangerous method of transport. The dread or fear invoked by the horrific scenes of two planes crashing into the Twin Towers of the World Trade Center in New York took over and led people to believe they were safer in a car. Years after the Fukushima nuclear power station meltdown, many residents still do not accept the assurances of scientists that levels of radioactivity are now perfectly safe. This isn't due to a difference in judgement about the risk but rather a result of the complete breakdown in trust between the public and experts caused by the disaster. Sometimes, people are very worried about risks that the available evidence says are incredibly unlikely while ignoring or accepting much more serious risks that we should protect ourselves against. Such risk perceptions aren't right or wrong; they just are. People aren't stupid or irrational – it's just that the formation of perceptions, attitudes and beliefs about risk is a complex business. Understanding the risk perception gap and how our views differ from established experts helps us make better judgements and contributes to our overall risk know-how.

Risk know-how means being informed and having a deep understanding of the risk and all the possible actions you could take. It is a kind of wisdom that doesn't increase anxiety but liberates us and allows us to focus on solutions. It is often best considered at a community level: while we might consume information and make sense of it as individuals, we tend to take action in groups and our collective understanding of risk and safety is what really matters. In 2022, the safety charity Lloyd's Register Foundation partnered with the campaigning organization Sense About Science and the Institute for the Public Understanding of Risk at the National University of Singapore to create a new initiative called Risk Know-How. They help local communities worldwide make sense of risk, given their context. For them, risk know-how means being able to ask questions, easily find reliable information, understand how the data can be manipulated, know what the magnitude of a risk is and how effective different responses might be and be able to make comparisons between the potential benefit or harm caused by acting or not acting. They also stress the importance of being able to handle uncertainty and how you can make good decisions when information is ambiguous or changing. Leonor Sierra, from , believes you know a community has risk know-how if 'they are not surprised by the consequences of a decision made regarding risk'.

Risk is all around us, whether crossing the road, sitting next to someone on the subway who might have a virus, or going out at night alone. It's exposure to danger and involves the possibility of loss or injury. While we have learned to live with risk and accept it, millions of people die every year from preventable accidents, so it isn't something we can afford to be complacent about. Risk refers to the future and 'the future exists only in the imagination'.[38] Safety, however, is about the here and now, the actions we take to protect people from the potential consequences of risk. But risk plays tricks on us, so we must understand it if we are to find solutions. In a rapidly changing world, risk is not constant. Rising temperatures due to climate change are dramatically increasing the threat of severe weather events such as tropical storms, drought and catastrophic flooding. Our world is also more interconnected than ever, meaning risks are more global, and the pace of change in emerging technology introduces risks such as cybercrime, AI and the misuse of our personal data. New risks are emerging all the time. Many of today's biggest challenges, such as nuclear war or global warming, were unheard of just one hundred years ago. We need to become more 'risk intelligent' as individuals and as a society to prepare for the unforeseen. Understanding risk, having the know-how, and being aware of how our brains work is critical to safeguarding our families and communities. It will allow us to become more resilient, survive dangers we can't yet imagine, and build a safer future for us all.

Notes

CHAPTER I

1 Reuters/Ipsos Poll: Russia/Ukraine Crisis W7 (Washington D.C., 6th October 2022).

2 Tom W. Smith, *Impact of Cuban Missile Crisis on American Public Opinion: Report of the John F. Kennedy Library and Museum* (16th October 2002).

3 *What Worries the World* survey (Ipsos, July 2023).

4 *The Global Risks Report* (World Economic Forum, 2023).

5 *Eurobarometer Survey* (European Parliament, Autumn 2022).

6 Alice Lee, Ian Sinha, Tammy Boyce, Jessica Allen, Peter Goldblatt. *Fuel poverty, cold homes and health inequalities.* (London: Institute of Health Equity, 2022).

7 The cost-of-living crisis is also a health crisis, *The Lancet Regional Health - Europe*, Volume 27, 2023, 100632.

8 Vilain, O. *Exclusive poll: Europeans on the brink* (Secours Populaire, Paris, 2022).

9 Hickman, Caroline et al., Climate anxiety in children and young people and their beliefs about government responses to climate change: a global survey. *The Lancet Planetary Health*, Volume 5, Issue 12, e863 - e873 (December, 2021).

10 John Aubrey, *Brief Lives,* pp. 272–4. (Clarendon Press, Oxford, 1696).

11 *Global report on drowning: preventing a leading killer.* (WHO, 2014.)

12 Gigerenzer G. Dread risk, September 11, and fatal traffic accidents. *Psychol Sci.* 2004 Apr;15(4):286-7.

13   Vincent van Gogh. *Letter to Theo van Gogh*, written 14 May 1882 in The Hague. Translated by Mrs. Johanna van Gogh-Bonger, edited by Robert Harrison, number 193.

14   Norman Lewis, *A Dragon Apparent* (Jonathan Cape, 1951).

15   Robert Macfarlane, *Mountains of the Mind: A History of a Fascination* (Granta, 2003).

CHAPTER 2

1   Guy D. Middleton, *Poseidon's Wrath* (Aeon, 2nd August 2021).

2   Pliny to Cornelius Tacitus from Pliny the Younger, *The Letters of the Younger Pliny* translated by Betty Radice (Penguin, 1963).

3   Homer, *Iliad*, IX.4-6.

4   W. H. Norton, 'The Influence of the Desert on Early Islam', *The Journal of Religion*, Vol. 4, No. 4, pp.383-396 (July, 1924).

5   Gaspar Mairal, 'The Mediterranean Origin of Risk', in *A Pre-Modern Cultural History of Risk* (Routledge, 2020).

6   Gaspar Mairal, 'When risk navigated to the Americas', in *A Pre-Modern Cultural History of Risk* (Routledge, 2020).

7   Extract from Cardano quoted in *Against the Gods: the Remarkable Story of Risk* by Peter Bernstein (Wiley, 1998).

8   John F. Ross, *Pascal's Legacy* (2004).

9   Giddens and Pierson, 1998.

CHAPTER 3

1   *Science*, 6 Dec 2013, Vol 342, Issue 6163, pp. 1214-1217.

2   Interview in *Time* magazine, 18th March 2011.

3   Richard Lloyd Parry, *Ghosts of The Tsunami* (Vintage, 2017).

4   Told to AP Newswire (6th March, 2011).

5   Tsujiuchi T, Yamaguchi M, Masuda K, Tsuchida M, Inomata T, Kumano H, Kikuchi Y, Augusterfer EF, Mollica RF. High

Prevalence of Post-Traumatic Stress Symptoms in Relation to Social Factors in Affected Population One Year after the Fukushima Nuclear Disaster. *PLoS One*. 2016 Mar 22;11(3).

6   Matthew Neidell, Shinsuke Ushida, Marcella Veronesi. Be Cautious with the Precautionary Principle: Evidence from Fukushima Daiichi Nuclear Accident, in *IZA Institute of Labour Economics* (October 2019).

CHAPTER 4

1   As of 16th November 2023.

2   Gideon Harvey, '*Morbus Anglicus or The Anatomy of Consumptions*', (London, 1666).

3   Glyn Taylor, '*How the allies won the war in 1918: strategic alignment or complete u-turn?*' King's College London, 7th June 2021.

4   Price GM. Influenza-destroyer and teacher: a general confession by the public health authorities. *Survey*. 1918; 41:367–9.

5   Science *Business*, 7th April 2020.

6   Taylor S, Asmundson GJG. Negative attitudes about face-masks during the COVID-19 pandemic: The dual importance of perceived ineffectiveness and psychological reactance. PLoS One. 2021 Feb 17.

7   Watson, Oliver J., et al., Global impact of the first year of COVID-19 vaccination: a mathematical modelling study. *The Lancet Infectious Diseases*, Volume 22, Issue 9, 1293 – 1302 (5th September 2023).

8   ABC News/Ipsos Poll, 11-12th September 2020.

9   American News Pathways March 2020 Survey, (15th April 2020, Pew Research Centre).

10   Coronavirus: US and UK Governments losing public trust, *The Conversation*, 6th May 2020.

11   Ahluwalia SC, Edelen MO, Qureshi N, Etchegaray JM. Trust in experts, not trust in national leadership, leads to greater uptake of recommended actions during the COVID-19

pandemic. *Risk Hazards Crisis Public Policy*. 2021 Sep.

12 Liu, J., Shahab, Y. and Hoque, H. (2022), Government Response Measures and Public Trust during the COVID-19 Pandemic: Evidence from Around the World. Brit J Manage, 33: 571-602.

CHAPTER 5

1 Robyn Spencer, Contested Terrain: The Mississippi Flood of 1927 and the Struggle to Control Black Labor, *The Journal of Negro History* 1994 79:2, 170-181.

2 A. Grinsted, P. Ditlevsen, & J.H. Christensen, Normalized US hurricane damage estimates using area of total destruction, 1900–2018, *Proc. Natl. Acad. Sci. U.S.A.* 116 (48) 23942-23946

3 Speech by George W. Bush, 31st August 2005 (George W. Bush Presidential Library).

4 *The New Orleans Hurricane Protection System: What Went Wrong and Why*, a report by the American Society of Civil Engineers Hurricane Katrina External Review Panel, 2007.

5 Keith Bea, Federal Emergency Management Policy Changes After Hurricane Katrina: A Summary of Statutory Provisions (Congressional Research Service, 6th March, 2007).

CHAPTER 6

1 Excerpts from Donald Trump's Victory Tour speech in West Allis, Wisconsin, 13th December 2016.

2 Joanne Hinds, Emma J. Williams, Adam N. Joinson, 'It wouldn't happen to me': Privacy concerns and perspectives following the Cambridge Analytica scandal, *International Journal of Human-Computer Studies*, Volume 143, 102498 (2020).

3 Lloyd's Register Foundation *World Risk Poll* 2019, powered by Gallup.

4 OECD, 2022.

5   The *Singapore Risk Barometer 2023*, Report of the LRF Institute for the Public Understanding of Risk at NUS.

6   ProPublica, 18th September 2018

7   Bartlett, Robert, et al. 'Consumer-lending discrimination in the FinTech era.' *Journal of Financial Economics* 143.1 (2022): 30-56.

8   Kröger, Jacob Leon, Milagros Miceli, and Florian Müller. 'How data can be used against people: A classification of personal data misuses.' *Available at SSRN 3887097* (2021).

9   Ipsos Poll conducted 1-7th April 2022.

10   *Americans and Privacy: Concerned, Confused and Feeling Lack of Control Over Their Personal Information* (Pew Research Center, November 2019).

11   *Olmstead vs. United States,* 277 U.S. 438 (1928).

12   Juan Pablo Carrascal, Christopher Riederer, Vijay Erramilli, Mauro Cherubini, and Rodrigo de Oliveira. 2013. Your browsing behavior for a big mac: economics of personal information online. In Proceedings of the 22nd international conference on World Wide Web (WWW '13). Association for Computing Machinery, New York, NY, USA, 189–200.

13   Ipsos Study conducted 1-7th April 2022 (as above).

14   Hanbyul Choi, Jonghwa Park, Yoonhyuk Jung (2018). The role of privacy fatigue in online privacy behavior, *Computers in Human Behavior*, Volume 81, 2018, Pages 42-51.

15   Thomas Friedman speaking at the Aspen Institute, 2nd August 2017.

CHAPTER 7

1   WHO Estimates of the Global Burden of Foodborne Diseases (2015, Switzerland: WHO).

2   OECD, 2018.

3   *Eating Animals* (2010, London: Penguin).

4   M. Satin, *Encyclopaedia of Food Safety*, Vol. 1, History of Foodborne Disease - Part 1 (Academic Press, 2013).

5   Francesco Berna et al., Microstratigraphic evidence of in situ fire in the Acheulean strata of Wonderwerk Cave, Northern Cape province, South Africa. *Proceedings of the National Academy of Sciences*, April 2, 2012.

6   Zohar, I. et al., Evidence for the cooking of fish 780,000 years ago at Gesher Benot Ya'aqov, Israel. *Nature Ecology and Evolution*, 2022 Dec;6(12).

7   R. Wrangham. *Catching Fire: How Cooking Made Us Human* (Profile, 2010).

8   Caporael, Linnda, R. *Science*, Vol. 192, Issue 4234, pp. 21-26 (2 Apr 1976).

9   Arrian, 7.26.1.

10 Tsouras, Peter G. *Alexander* (Brasseys, 2004).

11 Plutarch, 72.

12 *The Anabasis of Alexander*, Arrian. 401.

13 Oldach, D.W., Richard, R.E., Borza, E.N., & Benitez, R.M. (1998). *A Mysterious Death*. New England Journal of Medicine, 338(24), 1764-1769.

14 Grace, D. Burden of foodborne disease in low-income and middle-income countries and opportunities for scaling food safety interventions. Food Sec. 15, 1475–1488 (2023).

15 Sen, Amartya. *Poverty and Famines: An Essay on Entitlement and Deprivation* (Clarendon Press, Oxford, 1981).

16 *The Safe Food Imperative: Accelerating Progress in Low- and Middle- Income Countries* (World Bank Group, 2019).

17 Scarborough, P., Clark, M., Cobiac, L. et al. Vegans, vegetarians, fish-eaters and meat-eaters in the UK show discrepant environmental impacts. Nature Food 4, 565–574 (2023).

18 *The Seaweed Revolution*, Vincent Doumeizel (Hero, 2023).

19 *New Scientist*, 23rd December 1995.

20 Lord Phillips (chairman), *The BSE Inquiry* (London, Stationary Office, 2000).

CHAPTER 8

1   Erich C. Fisher et al., Coastal occupation and foraging during the last glacial maximum and early Holocene at Waterfall Bluff, eastern Pondoland, South Africa. *Quaternary Research*, (2020); 97:1.

2   UNEP, 2014b.

3   Edward B. Barbier, *Climate Change Impacts on Rural Poverty in Low-Elevation Coastal Zones*. Policy Research Working Paper 7475 (World Bank Group, November 2015).

4   NOAA (22nd August 2023).

5   Greene, C.A., Gardner, A.S., Wood, M. et al. Ubiquitous acceleration in Greenland Ice Sheet calving from 1985 to 2022. Nature 625, 523–528 (2024).

6   Simon Albert et al., 2016 Environ. Res. Lett. 11 054011.

7   *New Scientist*, 16th May 2016.

8   *Time* magazine, 11th November 2013.

9   E. E. Cummings, *95 Poems* (New York, 1958).

10 Homer, *The Odyssey* (Book 5, Lines 291–96.

11 Letter 676: To Theo van Gogh, Arles, Saturday 8th September 1888.

12 Timothy Judd, 'La Mer', Debussy's Sonic Portrait of the Sea (*The Listeners' Club*, 16th May 2018).

13 William Hamilton, Melville and the Sea in *Soundings: An Interdisciplinary Journal*, Winter 1979, Vol. 62, No. 4, pp. 417-429 (Penn State University Press).

14 Ernest Hemingway, *The Old Man and the Sea* (New York, Charles Scribner's Sons, 1952.

15 Joshua Slocum, *Sailing Alone Around the World* (Canada, 1900).

16 T. S. Eliot, *The Dry Salvages* (Faber, London, 1941).

17 Letter from John Doull, Fishery Office, Eyemouth 15th October 1881. National Records of Scotland reference: AF23/47 pp.169-170.

18 Daryl Attwood, *Insight report on safety in the fishing industry:*

a global safety challenge (Lloyd's Register Foundation, 2018).

19 S. Willis et al., The human cost of global fishing, *Marine Policy* 148 (2023).

20 Sam Willis and Eric Holliday, *Triggering Death: Quantifying the true human cost of global fishing* (Fish Safety Foundation, November 2022).

21 Nguyen, Quang & Leung, PingSun. (2009). Do Fishermen Have Different Attitudes Toward Risk? An Application of Prospect Theory to the Study of Vietnamese Fishermen. *Journal of Agricultural and Resource Economics*. 34. 518-538.

22 Mary E. Davis (2012), Perceptions of occupational risk by US commercial fishermen, *Marine Policy*, Volume 36, Issue 1, 2012, pp 28-33.

23 MSNBC News Service and Reuters, 10th September 2011.

24 Reuters, 12th September 2011.

25 Daryl Attwood, *Insight report on safety in the passenger ferry industry: a global safety challenge* (Lloyd's Register Foundation, 2018).

26 Report on the fact-finding missions to the Philippines (Ferrysafe, March and May, 2019).

27 Lincoln Paine, *The Sea and Civilization: A Maritime History of the World* (Random House, New York, 2013).

28 Marc Levinson, *The Box: How the Shipping Container Made the World Smaller and the World Economy Bigger* (Princeton, 2016).

29 S E Roberts and P B Marlow (2005), 'Traumatic work related mortality among seafarers employed in British merchant shipping, 1976 2002', *Occup Environ Med* 2005; 62:172–180.

30 Laleh Khalili, *Sinews of War and Trade: Shipping and Capitalism in the Arabian Peninsula* (London, 2020).

31 Nicolette Jones, *The Plimsoll Sensation: The Great Campaign to Save Lives at Sea* (London, 2006).

32 Thomas Thune Andersen, *If we want to meet the Paris Agreement, we must overhaul how we manage the ocean* (Climate Champions, 8th June 2021).

CHAPTER 9

1   *Global Status Report on Road Safety 2023* (World Health Organization).
2   National Safety Council, USA (2022).
3   Tom Vanderbilt, *Traffic: Why We Drive the Way We Do and What it Says About Us* (Allen Lane, 2008).
4   *Status Report for Road Safety in the WHO African Region 2023* (WHO, 2024).
5   'Africa sees sharp rise in road traffic deaths as motorbike taxis boom', *The Guardian*, 13th December 2023.
6   Ibid.
7   Paul Slovic, Baruch Fischhoff, and Sarah Lichtenstein, Accident Probabilities and Seat Belt Usage: a Psychological Perspective, *Accident Analysis and Prevention*, Vol. 10, (Elsevier, 1978).
8   Ibid.
9   Hyakawa, H. et al., Automobile risk perceptions and insurance-purchasing decisions in Japan and the United States, *Journal of Risk Research*, 3 (1), 51-67 (2000).
10 Peltzman, S. The effects of automobile safety regulation, *Journal of Political Economy*, 83, 677-725 (1975).
11 John Adams, *The Efficacy of Seat Belt Legislation* (UCL, 1981).
12 *NTSB chief to fed agency: stop using misleading statistic*, AP News, 19th January 2022.
13 William Kremer, *More than a million people die on roads every year: Meet the man determined to prevent them* (BBC Transport, 19th May 2024).

CHAPTER 10

1   Gillian D'Arcy Wood, *Tambora: The eruption that changed the world* (Princeton University Press, 2015).
2   Ibid.
3   Nick Bostrom, Existential Risks: Analyzing Human Extinction

Scenarios and Related Hazards, *Journal of Evolution and Technology*, Vol. 9 (March, 2002).

4   Toby Ord, *The Precipice: Existential Risk and the Future of Humanity* (Bloomsbury, London, 2020).

5   First reported in *Nature*, No. 1684, Vol. 65, (6th February 1902).

6   H. G. Wells, *The Discovery of the Future* (printed by B. W. Huebsch, New York, 1913).

7   *The Times*, 6th September 1859.

8   Nick Bostrom, Existential Risks: Analyzing Human Extinction Scenarios and Related Hazards, *Journal of Evolution and Technology*, Vol. 9 (March, 2002).

9   Nick Bostrom and Milan M Cirkovic, *Global Catastrophic Risks* (Oxford, 2008).

10   T. S. Eliot, *The Hollow Men* (Faber, 1925).

11   Nick Bostrom and Milan M Cirkovic, *Global Catastrophic Risks* (Oxford, 2008).

12   Martin Rees, *Our Final Century* (Arrow, 2004).

13   Toby Ord, *The Precipice: Existential Risk and the Future of Humanity* (Bloomsbury, 2020).

14   SJ Beard, Martin Rees, Catherine Richards and Clarissa Rios Rojas (eds), *The Era of Global Risk: An Introduction to Existential Risk Studies*. Cambridge, UK: Open Book Publishers, 2023.

15   World Economic Forum, *Global Risks 2006* (2006).

16   World Economic Forum, The Global Risks Report 2024 (WEF, 2024).

17   United Nations Office for Disaster Risk Reduction, *Global Assessment Report on Disaster Risk Reduction 2022: Our World at Risk: Transforming Governance for a Resilient Future*. (Geneva, 2022).

18   Marcus Tullius Cicero, *For Sextus Roscius of Ameria* (29.79).

19   D. E. Alexander, Resilience and disaster risk reduction. an etymological journey, *Natural Hazards and Earth System Sciences*, 13, 2707-2716 (2013).

20   Michael Bruno, *Foresight Review of Resilience Engineering* (Lloyd's Register Foundation, 2015).

21   European Commission, 2016.

22   The UN Food and Agriculture Organization (FAO) definition of resilience: 'the ability to prevent disasters and crises, as well as to anticipate, absorb, accommodate or recover from them in a timely, efficient, and sustainable manner.' (FAO, 2013.)

23   Ernest Hemingway, *A Farewell to Arms* (New York, Scribner, 1957).

24   *After the Great East Japan Earthquake: a review of community engagement activities and initiatives* (Nuffield Council on Bioethics, 2019).

25   *Stories of resilience in Türkiye: Rebuilding livelihoods after the earthquakes* (IFRC Website Article, 18th June, 2024).

26   Lloyd's Register Foundation World Risk Poll 2023, powered by Gallup.

27   Steven M. Southwick, Dennis S. Charney, and Jonathan M. DePierro. *Resilience: The Science of Mastering Life's Greatest Challenges* (Cambridge University Press, 2023).

28   Carolyn J. Lo, Leonard Lee, Weichang Yu, E Shyong Tai, Tong Wei Yew, & Isabel L. Ding. Mindsets and self-efficacy beliefs among individuals with type 2 diabetes. *Scientific Reports* 13, 20383 (2023).

29   Carol Dweck, *Mindset: The New Psychology of Success* (Random House, 2006).

30   Daniel P. Aldrich, *Building Resilience: Social Capital in Post-Disaster Recovery* (Chicago The University of Chicago Press, 2012).

31   Bonfanti RC, Oberti B, Ravazzoli E, Rinaldi A, Ruggieri S, Schimmenti A. The Role of Trust in Disaster Risk Reduction: A Critical Review. *International Journal of Environmental Research and Public Health*. 21(1):29 (2023 Dec 24).

32   Robert D. Putnam, *Bowling Alone: The Collapse and Revival of American Community* (Simon and Schuster, New York, 2000).

33 Ando, R. (2018). Trust—What Connects Science to Daily Life. *Health Physics*, 115 (5), 581-589.

34 Amos Tversky and Daniel Kahneman. Judgment under Uncertainty: Heuristics and Biases. *Science* 185,1124-1131 (1974).

35 B. Fischhoff, P. Slovic, S. Lichtenstein, S. Read, and B. Combs, How Safe is Safe Enough? A Psychometric Study of Attitudes Towards Technological Risks and Benefits, *Policy Sciences*, 9, pp.127-152 (Kluwer, 1978).

36 P. Slovic, Perception of risk. *Science*, 236, pp. 280-205 (1987).

37 Ibid.

38 John Adams. Not 100% sure? The Public Understanding of Risk in *Successful Science Communication: Telling It Like It Is* (Cambridge University Press, 2011).

Acknowledgements

In any major project, motivation and inspiration are essential. Motivation for attempting this book came from another Hero author (and good colleague), Vincent Doumeizel: I wanted to attempt to do for risk what he did to raise public awareness of the importance of seaweed in *The Seaweed Revolution*. Inspiration came from my 'dear friend' and pioneer of creative risk communications, Pablo Suarez, who sadly passed away while I was writing this book. Pablo approached everything he did with love, kindness and humour. He believed that we will only succeed in communicating risk to local communities if we connect with people at the human level. One way to do that is through good, honest storytelling, and I hope this book lives up to Pablo's legacy.

I am also indebted to Hania Farhan from Gallup, whose deep knowledge of the existing risk literature and what we know about public attitudes to risk was a great help in developing the concepts and themes that make up this book.

Some chapters benefited from valuable feedback and guidance kindly provided by subject experts, including Mariko Nishizawa on public perception of nuclear power in Japan, Olivia Swift on ocean risk and the maritime industry, Daryl Attwood on the dangers associated with fishing and ferries, and Leonor Sierra on Risk Know-How. I would also like to thank Roy and Lesley Adkins for their helpful guidance on maritime history and the publishing process.

My colleagues at Lloyd's Register Foundation and the LRF Institute for the Public Understanding of Risk at the National University of Singapore have been an incredible source of ideas, advice and support for which I am extremely grateful. Special thanks are due to Professor Richard Clegg, Ruth Boumphrey, Tim Slingsby, Louise Sanger, Beth Elliot, Samuel Dadd, Professor Leonard Lee, Olivia Jensen, Yiyun Shou, Trinh Thi Tra, Jared Ng, Reuben Ng and Chia-Wen Wang. My thanks also go to the *World Risk Poll* team, especially Andy Rzepa, Andrew Duggan, Benedict Viggers, Ed Morrow, and Aaron Gardner, for help translating data from the poll and putting it into a real-world context.

Throughout the writing of this manuscript, I was lucky enough to have two unpaid assistants who allowed me to test each chapter and were honest with me about how interesting the content was, helped rein in my more outlandish flights of fancy, and corrected the worst of my mistakes. If this has become a readable and engaging book, it is largely thanks to the help of Emma Ackerman and Sam Young.

I would also like to thank Tom Chalmers and the wonderful team at Hero, especially my editor, Christian Müller, for their hard work, helpful guidance, and belief in the need for a popular book about the importance of understanding risk. I hope it is justified and that this book goes some way to making the world a safer place.

Index